D1060627

Practical Environmental Bioremediation

The Field Guide

Practical Environmental Bioremediation

The Field Guide

R. Barry King
Gilbert M. Long
John K. Sheldon

Lewis Publishers
Boca Raton Boston London New York Washington, D.C.

Library of Congress Cataloging-in-Publication Data

King, R. Barry.
Practical environmental bioremediation / R. Barry King, Gilbert M. Long, John K. Sheldon. -- 2nd ed.
p. cm.
Includes bibliographical references and index.
ISBN 1-56670-208-9 (alk. paper)
1. Bioremediation--Handbooks, manuals, etc. I. Long, Gilbert M. II. Sheldon, John K. III. Title.
TD192.5.K56 1997
628.5--dc21 97-37142
CIP

Direct all inquiries to CRC Press LLC, 2000 Corporate Blvd., N.W., Boca Raton, Florida 33431.

Lewis Publishers is an imprint of CRC Press LLC

International Standard Book Number 1-56670-208-9
Library of Congress Card Number 97-37142
Printed in the United States of America 1 2 3 4 5 6 7 8 9 0
Printed on acid-free paper

Frontispiece

"To the modern student of Ecology and Environment, it is instructive that all artificial pollution events, both regional and global, find their origin in ***MAN*** *and his activities and, at the same time, the major modes of natural cleansing are accomplished by* ***MICROBES*** *through their activities; on the one hand, God's highest achievement in Creation is found doing the most destruction, while the lowliest life-form crafted by His hand does the janitor's duty — the former by disobedience and the latter by Design."*

(author unknown)

Preface to the First Edition
Why Bio?

Bioremediation, or enhanced microbiological treatment, of environments contaminated with a variety of organic and inorganic compounds is one of the most effective innovative technologies to come along in this century. Actually, it is simply a new application of a very old technology once primarily used in wastewater treatment. Today, bioremediation is routinely applied to soils, sludges, groundwater, process water, and surface waters contaminated with such chemicals as crude oil, petroleum hydrocarbons, fuels, industrial solvents, and wood-treating agents.

Authors John Naisbitt and Patricia Aburdene, in their visionary best-seller *MEGATRENDS 2000*, predicted that increased applications in biotechnology would be one of the nation's major new directions during the 1990s, in effect dubbing the decade "The Age of Biology".[1] The Hazardous Materials Control Research Institute (HMCRI), reporting in March 1990 on an assessment of the environmental management market, titled their article "Hazardous Waste Bioremediation Targeted as Fastest Growing Market in Environmental Management".[2] The HMCRI hosts an annual international meeting addressing bioremediation in Washington, D.C. Professor Dave Gibson, speaking at the first annual HMCRI bioremediation conference in 1988, said that it was his opinion that, by the end of the century, the majority of hydrocarbon-contaminated sites would be remediated by employing some form of biological treatment.[3]

Bioremediation has been successfully applied to enhance natural degradation of petroleum hydrocarbon and organic chemicals at sites throughout the world. Gasoline and diesel contamination is especially responsive to this technology, both in soils and groundwater. Even heavy fuels and crude oil have been degraded in a timely fashion through the use of microorganisms.

The bioremediation activities on the oily beaches of Prince William Sound caused by the Exxon Valdez spill in 1989[4] were widely publicized. Similar technology was applied to crude oil spills along the Texas Gulf Coast during 1990.[5,6] In a meeting called by the U.S. Environmental Protection Agency (EPA) (February 22, 1990) to discuss environmental applications of biotechnology, Administrator William Reilly reiterated his support for the development and application of alternative technologies and, specifically, bioremediation. He added that the application of biological treatment might be the breakthrough that is needed to offset the staggering costs of compliance with Superfund.[7]

The Superfund Amendments and Reauthorization Act of 1986 (SARA) emphasizes treatment over containment to remediate Superfund sites. In addition to ensuring protection of human health and environment, two "threshold criteria" for selecting remedies under proposed revisions to the National Contingency Plan are that they provide long-term effectiveness and achieve the greatest reduction in toxicity, mobility, and volume. This amounts to a remedy that provides treatment to the maximum extent practicable.[8] When complete destruction of wastes can be achieved, both of these criteria are met by employment of bioremediation. In addition, the costs of remediation through biological treatment are attractive when compared with other technologies.

In the past, contaminated soils and sludges were either burned, buried, or chemically treated in place. These remedial alternatives are often costly and regulatory compliance is difficult. Incineration is both costly and increasingly harder to implement. Landfill and *in situ* fixation do not destroy waste, and landfilling only changes the place of residence, delaying future liability. Buried liability often does not remain hidden, and solidified waste may simply leach out at a slower rate. Innovative bioremediation technologies may offer complete destruction of contaminants and can often be applied at a lower total cost, at a faster rate, and with no lingering liability.

Biological treatment technologies are finding widespread public support as an "organic" approach to site reclamation. Public hearings concerning proposed use of biotechnology for site remediations have not met with the criticism and outcry that other alternatives have had to face.[7] Bioremediation uses *natural* biological processes to clean up the chemical contamination caused by man's activities. The fact is, whenever organic chemicals and other hydrocarbons are spilled or otherwise introduced into the environment, naturally-occurring microbes immediately begin to acclimate themselves and biodegrade this new food source. The reasoning suggests, "If it happens anyway, why not help it along the best we can?"

The EPA believes so strongly in the future usefulness of bioremediation and other innovative technologies for treating contaminated soils and groundwater that it created a special Technology Innovation Office in August 1990 to encourage and focus on these new technologies.[9]

There has been a long-standing need for a definitive book on bioremediation of contaminated environments. To date (1991), no comprehensive handbook has appeared solely addressing the fundamentals of field application of bioremediation. Contributing to this lack of sufficient texts on this subject may be the highly fragmented nature of the industry, corporate indecision to fully fund and support such a new technology, the hesitation of regulators to issue permits for techniques they don't thoroughly understand, and the natural reluctance of bioremediation contractors to publish data on techniques they believe are proprietary. The reality is that there are no secrets or unique concepts for successful application. There is simply a well-defined set of prerequisites that must be understood, intelligently addressed, and carefully applied to obtain a successfully completed project.

This *PRACTICAL ENVIRONMENTAL BIOREMEDIATION* is written with the benefit of the authors' combined 55 years of environmental management and restoration experience. The authors have worked together for several years actively applying bioremediation technology in the field. Specific and general experience gained from actual field projects where the authors and others have successfully applied this technology are presented in the text.

Dozens of projects have been completed and closures obtained in many geographic settings in the United States, attesting to the expanding role of bioremediation as a viable and cost-effective remedial alternative. It is a fitting tribute indeed to the effectiveness and economics of this blooming technology that research efforts are being funded by industry, government, foundations, and major universities around the world to explore wider applications for bioremediation as a means of environmental restoration.

We believe this book will be of practical value to students, consultants, design engineers, managers, regulators, scientists, and others involved in the environmental sciences and restoration activities. This book focuses on technology that is available today. Incorporated into the text are practical examples familiar to the authors. The concepts presented have been tested and demonstrated successfully in dozens of applications for a variety of clients and have become the basis for seminars, workshops, short courses, and university lectures given by the authors. These concepts and remediation techniques can be immediately and effectively applied if the principles described herein are thoroughly understood and a well organized project plan is constructed, appropriately conducted, and the process is carefully monitored.

REFERENCES

1. Naisbitt, J. and Aburdene, P., 1990, *Megatrends 2000*, Chapter 8, The Age of Biology, Wm. Morrow, New York, NY.
2. Anon., *HMCRI Focus*, 1990, Hazardous Materials Control Research Institute, Silver Spring, MD, March, p. 10.
3. Gibson, D.T. Professor of Microbiology, University of Iowa.
4. Anon., Alaska's big spill; can the wilderness heal?", *National Geographic*, January 1990, 177, (1), 5–43.
5. MEGA BORG Oil Spill Off the Texas Coast: An Open Water Bioremediation Test, *Combined General Land Office & Water Commission Report*, State of Texas, Austin, July 12, 1990, 13 pp. and Appendix.
6. Combatting Oil Spills Along the Texas Coast: A Report on the Effects of Bioremediation, *General Land Office Report*, State of Texas, Austin, June 12, 1990, 14 pp. and three Appendices.
7. *Environment Reporter*, Bureau of National Affairs, Washington, D.C., March 2, 1990, p. 1820.
8. Revisions to NCP weight criteria for site remedies, *Waste Tech News*, February 26, 1990, p. 3.
9. Anon., New EPA office encourages use of innovative treatment technologies, *HMCRI Focus*, August 1990, p. 4.

Preface to the Field Guide

The publishing of the first edition of this book, "Practical Environmental Bioremediation", in 1992 marked the advent of the first effective *field guide* to application of this innovative remediation technology. Previous volumes by others, although adding significantly to the knowledge base in this technology, consisted mainly of complex engineering manuals, handbooks of general data derived from the literature, and compendiums composed of various collections of research papers. Most of this information had been published elsewhere, or presented as short topics within larger volumes addressing cleanup of specific media.

This book sets out primarily to condense the knowledge base we have at hand (drawing on the existing data base referenced above) and to present the necessary technical aspects and concepts in language that can be readily comprehended by the general public, the technical student, and the aspiring newcomer to this exciting natural cleanup technique. The brisk sales of the first edition, the enthusiastic response from readers, and the adoption of the book as a text in several universities and technical schools were strong encouragement to the authors. In the 5 years since the original edition appeared, those of us working in this technology have witnessed great strides toward simplifying treatability formats, finding new approaches to field application, more potent nutrient formulations, monitoring protocols, and the resulting general improvement in results. Several new bioremediation texts have recently appeared addressing specific segments of the environment via specific remedial techniques. All this newly published information and field experience demanded a second look at this emerging technology and a new edition of this book.

Recent publications have greatly increased the information available to practitioners and site owners as they attempt to intelligently select effective remedial treatment alternatives. Laboratory and field research have documented new areas for concern during site characterization and new avenues for full field application of this innovative technology (*Handbook of Bioremediation,* Lewis, 1994). In addition to reporting these new findings, actual site closures and new case histories are presented in this new edition as they have recently become available from the authors' personal experiences, technology transfer conferences, and the literature. In all, there has been at least a doubling of verified available knowledge in applied bioremediation technique since the first edition of this book was published in 1992.

In this updated and expanded field guide, chapter titles and headings have changed as appropriate, and illustrations have been updated or added for clarity. New material has been added to reflect the modern image of this fast-emerging industry, and case histories are provided to illustrate and amplify understanding of specific applications. Whole new chapters have been added to cover new developments in designing the Bioremediation Laboratory, Lagoon Bioremediation, *In Situ* Aquifer, and Vadose Zone Bioremediation, and new information has been added on the techniques for remediation of chlorinated solvents and mixed waste. Besides these several organizational changes, every chapter has been thoroughly revised and updated. Each sentence has been reviewed for clarity and changes made where necessary. The authors' relatively simple, direct writing styles were taken as an effort toward readability. The researcher in pure science and many advanced graduate degree candidates may find the material less than stimulating, as there is a decided lack (indeed a conscious deliberate avoidance) of highly technical terms, convoluted mathematical equations for reaction kinetics, and advanced biochemistry that might prove an impediment or superfluous, rather than a benefit, to a clear understanding of the basics of bioremediation. In keeping with this, many reader responses have been incorporated into the field guide, particularly in the area of expanding certain sections for greater clarity and providing more basic information for the student, geologist, and the engineer who may be less familiar with concepts in microbiology and biochemistry.

At length, environmental microbiology, in its direct practical application for bioremediation of contaminated sites, is an exceptionally broad discipline encompassing biochemistry, microbial ecology, geochemistry, and the basics of reaction kinetics. It requires input from, and the considered appreciation of, a team of geologists, hydrologists, chemists, microbiologists, engineers, and field technicians to ensure the best chance for successful application. Discussion of these topics has been approached with an eye toward simplicity, since in the authors' view, it is much more important to accomplish a simple reduction of site contaminant concentrations to below action limits than it is to identify specific *genus and species* of organisms doing the work or delineating precise biochemical degradation pathways.

This new field guide can be used effectively in teaching a thorough short course in Bioremediation Basics, especially when used in combination with the full-color slides, overheads, and instructor's notes that have been prepared from the illustrations found throughout the book. Entry into the field of bioremediation can be expedited and many of the common mistakes commited by novice "wanna-bee" practitioners can be avoided through use of these materials, which present the basics that must be understood for successful application of bioremediation in the field.

Address inquiries to:

Bioremediation Basics
P.O. Box 7595
Albuquerque, NM 87194

R. Barry King

P.O. Box 7595 ABQ, NM 87194 — (505)836-6044 — bking@TVI.cc.nm.us

R. Barry King is an environmental consultant possessing over 33 years of experience in civil engineering, pollution control, and remediation of contaminated environments. He has worked for major remediation firms, private industry, railroads, banks, and foreign and domestic governments as a field expert in application of remedial techniques for residual hydrocarbons, wastewater treatment, leaking underground storage tanks, biodegradation of radioactive mixed waste, fate of contaminants in geologic storage, and the various techniques for bioremediation of soils, sludges, and groundwater. He has been involved in the conception, design, construction, and operation of waste treatment systems for resource recovery and delisting of RCRA K-category wastes.

Mr. King has authored many EPA and DOE reports on monitoring, degradation, control, and treatment technologies for hazardous and mixed waste. He has written numerous SuperFund and RCRA site work plans for bioremediation as the primary means of closure. He is a professor of advanced courses in Environmental Technology at the Albuquerque Technical Vocational Institute (TVI) in New Mexico. He conducts remediation seminars for various regulatory agencies nationwide and has taught the bioremediation section of the California State Certification in Hazardous Materials Management (CHMM) course at the University of California Extension, Santa Cruz. He hosted the 1995–1996 EPA Bioremediation Technology Transfer Conferences on a grant from the Western Region Hazardous Substances Research Centers at Stanford and Oregon State Universities. In addition, he lectures in field bioremediation practice for the Waste Management Education and Research Consortium (WERC) at New Mexico State University, Las Cruces.

Mr. King has served the U.S. Department of Energy as a consultant under contract to provide guidance on Long-Term Performance Assessment of nuclear waste geologic disposal at the Waste Isolation Pilot Plant (WIPP) in New Mexico, and also for the Engineered Alternatives Task Force on transuranic radioactive mixed waste forms to be entered at the WIPP. He conducted the first underground microbiological survey of the WIPP in 1989. He has performed analytical evaluations of microbial activity that can be expected in the underground environment at the proposed Yucca Mountain Nevada High-Level Nuclear Waste Repository.

Mr. King earned a B.S. degree in Microbiology and M.S. in Environmental Biology from the University of Houston, Texas. While attending the University, he worked full time for Harris County Pollution Control Department under the venerable Walter Quebedeaux, Jr., Ph.D., J.D., who was his much revered mentor. He is a member of the American Society for Microbiology, Association of Ground Water Scientists and Engineers, and Hazardous Materials Control Research Institute. He works, teaches, writes, and lives in Albuquerque, NM.

Gilbert M. Long

Gil Long's area of expertise in bioremediation is design and system operations. After 15 years in industry as a process engineer, project engineer, and plant engineer in a variety of chemical industries, he joined a bioremediation development group as a project engineer. He has since been a project manager for numerous bioremediation projects using several of the methodologies described in this book.

Mr. Long has designed, installed, and/or managed the process monitoring for four full-scale and one pilot *in situ* bioremediation systems, three full-scale and one pilot land treatment cells, five full-scale and numerous pilot soil vapor extraction systems, one vadose zone bioremediation, one biovault demonstration project, and one lagoon bioremediation project. These projects involved regulatory interactions and negotiations with federal, state, and local agencies in developing detailed work plans and obtaining permits. Of the full-scale remediation projects, all but one have been successfully closed and one is still pending.

He has a Bachelor of Science in Chemical Engineering from Lafayette College and a Master of Science in Chemical Engineering from the University of Massachusetts. He is also a Certified Hazardous Materials Manager. His trade association memberships include the American Institute of Chemical Engineers (AIChE) and the Air & Waste Management Association (AWMA). He has served as a consultant on remediation technologies for the Synthetic Organic Chemicals Manufacturing Association (SOCMA).

Mr. Long has authored five papers for technical conferences and co-chaired a session at another. He is presently a Senior Program Manager for Remediation Technologies for Philip Services in Houston, TX.

Address:

Philip Services Corporation
1440 Sens Road
La Porte, TX 77571
(281)470-1388
(281)470-1399 (fax)
e-mail: gil-Long@philip-serv.com

John K. (Jack) Sheldon

John K. (Jack) Sheldon has 16 years of experience in the field of biotechnology gained in the paint and coatings, specialty chemicals, and environmental industries. For the last 11 years, he has been applying bioremediation technologies to sites worldwide often coupling those technologies with other physical/chemical processes.

Presently, Mr. Sheldon is a Principal Environmental Scientist — Innovative Technologies with Montgomery Watson of Des Moines, Iowa, and leads the company's bioremediation effort. During his years in the industry, Mr. Sheldon has been employed as a laboratory manager, marketing director, operations manager, project manager and program director. These positions have allowed him an opportunity to view the industry from the key functional perspectives. His experiences are relayed in this book.

Mr. Sheldon has been involved in installation of many of the systems described in this book. He has also been fortunate to have worked on 12 first-of-their-kind bioremediation systems in different states. Currently, he is actively working with bioremediation technologies focused on pesticides, explosives and coal tar wastes. He is also working on several natural attenuation projects involving petroleum hydrocarbons and chlorinated solvents.

Mr. Sheldon has authored 12 technical papers and has given more than 50 major presentations to industry and government clients and various technical associations. He has given presentations to more than half the country's regulatory agencies and has lectured in the Corrective Action Continuing Education courses at the University of Wisconsin and the University of Minnesota. Mr. Sheldon has presented a symposium on bioremediation at a recent American Society for Microbiology annual meeting. He has been a keynote speaker at the annual Environmental Professionals of Iowa technical meeting and at an annual worldwide environmental meeting for a major pharmaceutical company.

Mr. Sheldon earned a B.S. in Bacteriology and Public Health and a Masters degree in Industrial Microbiology from Wagner College on Staten Island, New York. He is a member of the American Society for Microbiology, the Society for Industrial Microbiology, the National Groundwater Association, and the American Society for Testing and Materials (Committee E-50-environmental Assessment). He has been a reviewer for the National Science Foundation, CRC Press, and other major publishers.

Acknowledgments

The authors wish to acknowledge the significant contribution of Dr. H. Eric Nuttall, Department of Chemical and Nuclear Engineering, Univ. of New Mexico, Albuquerque, whose liberal donation of time and effort enlarged the sections on bioremediation of metals and radionuclides.

R. B. K. wishes to express particular thanks to the late Walter A. (Doc) Quebedeaux, Jr., Ph.D., J. D. He was a true friend and a champion for a better environment for his fellow Texans and the world, and to the Troops at Harris County Pollution Control for their friendship and support. To Nick Fotheringham, Ph.D., formerly of Scripps Institute and the University of Houston, under whose tutelage I was able to earn an advanced degree. To my family for their unswerving support. To Roger Hansen, Esq., of Denver, CO, my dear friend, for review and corrections to the Preface and Chapter 1, to ensure that my history was correct.

G. L. wishes to express particular thanks to Dr. Robert B. Long, who read the manuscript to make sure we had not lost the audience, for his support of this and many other ventures. To my family for their support.

J. S. wishes to express particular thanks to his loving family, Gail, Andrew, and Kevin for their never ending support. To my parents because they have always been there for me. To special people like Mary, Larry, Betty, and Walter, who have inspired me. To my mentors, Scott Gonge and John Hansen, for setting my career on an upward course.

Contents

CHAPTER 1

Introduction: A Historical Perspective

The use and manipulation of microorganisms for treatment of municipal wastewater have been applied since ancient times. The Romans and others built intricate networks of sewers as early as 600 B.C. for collecting wastewater which underwent subsequent biological treatment. Though this phenomenon was only vaguely perceived, Roman architects planned, designed, and constructed sewage systems[1] that served their intended purpose for centuries. Collection vats and lagoons were provided for prevention of system backup and overload. These were the sites where microorganisms did the work of biodegradation of organic waste. What the Romans saw as a sort of self-purification took place as a function of retention time through the action of microbial digestion of the organic waste.

Sewage systems of modern design still operate in a very similar manner. Microbial degradation was found to proceed at a much faster pace when aided by aeration to provide oxygen for *aerobic* processes, though the *anaerobic* degradation process remains well suited for sludge digestion and certain bioreactor applications. Industrial and agricultural fermentations carried out by microbes produce a major portion of the world's feed silage and account for *all* the production of ethanol spirits consumed by humans. Although industrial ethanol feedstocks are produced in chemical reactors, the drug and pharmaceutical producers frequently use microbial transformations. The total annual dollar revenue generated through biotechnology in the U.S. is in the tens of billions.

For our present practical discussion, the concept of environmental bioremediation is "*A treatability technology that uses biological activity to reduce the concentration or toxicity of a pollutant. It commonly uses processes by which microorganisms transform or degrade chemicals in the environment.*"[2]

REGULATORY HISTORY

Historically, federal environmental legislation in the U.S. began with the **Rivers and Harbors Act** of 1899. Hazardous materials in the environment began to be addressed with enactment of the **Insecticide Act** of 1910, followed not so closely by the **Federal Insecticide, Fungicide, and Rodenticide Act** of 1947, and the **Solid Waste Disposal Act** and **Water Quality Act** of 1965. During the 1960s, those of us who were working on the enforcement side soon found the inadequacy of these early federal attempts to control pollution. Accordingly, several states (notably Texas, California, and New Jersey) began to pass individual state Clean Water and Clean Air Acts. At first, the remedies at law under these statutes in Texas were applicable in the civil courts only (making a successful prosecution by a Pollution Control Authority very difficult). As these early regulations were amended to allow criminal prosecution at the state and county levels, cases could be swiftly heard and the environment (and society) began to be more adequately protected.

With passage of the federal **National Environmental Policy Act** (1969), **Clean Water Act** (1972), and **Safe Drinking Water Act** (1974), and creation of the Environmental Protection Agency in late 1970, we finally had a more realistic means of correcting misuse and abuse of our land and water resources nationwide. In 1976, the **Resource Conservation and Recovery Act** and the **Toxic Substances Control Act** (RCRA and TSCA) brought soil and water resources under the regulatory microscope in a context that is both comprehensive and restrictive. **SuperFund** (CERCLA 1980) and the **Amendments and Re-authorization Act** (SARA, 1986) created a focus on costs and permanent solutions to contaminated sites. Site owners and waste generators were put on notice that contamination created by past activities would now have to be remediated. The specter of lingering liability could no longer be tolerated as the law provided no statute of limitations that could excuse the environmental sins of the past.

As more progressive regulations hit the books, a more stringent view of our environmental future was forced. Industry and (now) government alike were encouraged to do better — *much* better. Suddenly we became husbands to the earth and were informed that we'd have to start cleaning up the messes we'd already made, as well as minimizing any future impacts. The heyday of the standard technologies of burning or burying pollution was coming to an end. Effective means of pollutant destruction were needed at attractive prices to offset the increasing difficulties with landfilling and incineration.

During the two decades prior to promulgation of initial regulations under **RCRA** and **CERCLA**, the major focus of regulatory attention in the area of environmental management was on preservation of our atmospheric and surface water resources, wildlife and natural habitats, and abatement of detrimental activities affecting them. It was during this time that the authors and many present-day environmental professionals were gaining their early experience in environmental control. It was a time of air and water regulations, **NPDES** effluent discharge permits, and the early awakening of *real* concern for our natural resources. It was the **"Age of Pollution Control".**

Hazardous wastes, SuperFund, and lists of "priority pollutants" were still in the future. Immediate health effects and public nuisances were the governing parameters by which the activities of private parties and industry were policed; the government had not yet perceived any good reason to police itself. Generally, what was practiced within a property boundary was of little concern to the public or a neighbor, unless the activity produced some immediate nuisance effect.

Waste dumping, chemical spills, and atmospheric smoke or odors usually resulted in a citizen complaint (industry wasn't especially prone to complain about its industrial brothers at the time) that suggested a possible violation of a local air or water act, or simply constituted a common law nuisance. Complaints brought investigations that sometimes resulted in charges by local authorities, criminal or civil prosecutions, penalty assessments, restraining orders, cease and desist mandates, writs of Mandamus, an occasional jail sentence, and ultimate cessation of the offending activity. Any environmental damage that had been done to "public resources" was dismissed by most people as someone else's problem, if it was considered at all. The offender having been caught and punished, "all was well in the world", or so we thought.

With the passage of the National Environmental Policy Act (NEPA) in 1969 and subsequent environmental legislation, a more responsible view of our environmental future was stimulated. The realization that terrestrial, atmospheric, and aquatic resources might, in fact, be finite led some early visionaries to encourage legislation that required cleanup of this "mess" and prophylactic measures to further inhibit degeneration of our natural resources. We were moving steadily toward the **"Age of Remediation".**

With the implementation of RCRA beginning in 1980 and the passage of Super-Fund, greater emphasis was placed on prevention and remedial actions in the area of waste management. Waste minimization, hazardous materials control, and site remediation are the order of the day at present, and still tougher regulations are guaranteed. Hazardous waste treatment and safe disposal are primary national concerns. Management of newly generated waste and the need for remediation of existing contaminated sites have created a vast market that will stretch over at least the next several decades. The U.S. Department of Energy is working under a Thirty-Year Plan for cleanup that is slated to cost hundreds of billions of dollars by the year 2019 for DOE sites alone.[3]

Fueling this unprecedented period of environmental concern are the driving forces of increasing public awareness, development of more stringent regulations, and concern by waste generators over potential long-term environmental liabilities. As regulatory strategies continue to evolve, several trends become obvious: more waste constituents are being declared hazardous, waste volumes are increasing, and expensive, high-tech treatment options and Best-Demonstrated Available Technologies (BDATs) are being required.

The RCRA requirements for predisposal treatment (**Land Ban** disposal restrictions) and the new CERCLA/SARA requirements are forcing a greater emphasis on waste minimization, resource recovery, recycling, on-site treatment/remediation, and, ultimately, complete destruction of the waste or its toxicity. Reliance on innovative on-site and *in situ* treatments which can be contracted to hazardous waste management companies will free many waste generators from the economic drain of internal waste management. Many companies generating wastes want to focus their efforts on their primary business and leave waste management, treatment, and disposal to qualified consultants and contractors.

The federal time limit for pollution liability is clearly defined under RCRA as lasting *"from the cradle to the grave"*. There is *NO statute of limitations* for pollution liability. Also, just as clear is the notion that *"once an RCRA waste, always an RCRA waste"* and this follows all the way to final destruction or disposal. It is common knowledge that a waste generator remains responsible for the fate of generated waste even though it has been delivered to others for treatment and disposal. Thus was born the term "Potentially Responsible Party" (PRP) under CERCLA. It is simply a fact of life in U.S. industry that *"as long as the waste EXISTS, liability PERSISTS."*

It is this specter of continuing liability that constitutes one of the main reasons bioremediation is becoming such an attractive treatment alternative for hazardous and toxic wastes. Simply put, the major benefits of this technology include:

- Attractive economics
- Undisturbed environment (with *in situ* application)
- Universal (can often treat water, soil, sludges, air)
- Contaminants are destroyed or detoxified
- Liability is eliminated

Even if there is only an even *(50–50)* chance of success when using bioremediation as opposed to some other alternative, it is usually worth a try. *A bioremediation approach may not be right for every site, but when it's right, it's wonderful.* The most that can be at stake if bioremediation proves ineffective is possibly 2 months of time and several thousand dollars in expense. On the other hand, if the technology proves acceptable for the particular site in question, it can potentially save a large percentage of the cost of some other treatment technology.

EARLY SUCCESSES

While most pollution events were left with no attempted cleanup forthcoming, the atmosphere and most surface waters could undergo a sort of self-remediation through dispersion and natural cleansing once the pollution sources were controlled or eliminated. Not so for many sites with contaminated surface and subsurface soils and groundwaters which often remain a continuing threat until effective remediation can be applied.

Early attempts at remediation of sites contaminated with hydrocarbons, metals, and organic chemicals included the more common physical and chemical separations and reactions that were "proven and ready means" of dealing with immediate problems. Many of these methods simply removed the contamination from one medium and placed it in another. Aside from incineration (at horrendous expense), these methods were nondestructive and left the generator with continuing liability.

What was needed was a convenient method that would allow the destruction or detoxification of these unwanted contaminants in a manner similar to the natural healing in the atmosphere, rivers, and streams. For air pollution, this consisted of fallout, rainout, and *natural oxidation* (with energy supplied by the sun). For streams, it was *natural oxidation* enhanced by the energy supplied via resident microbes (or natural bioremediation). At that time, no one realized precisely what was taking place or understood that it could be applied for cleanup of contaminated soils and groundwaters.

It had been well known to petroleum refiners that API separator sludge and other oily wastes, when placed upon the ground across the "back 40 acres," could be seen to undergo a slow degradation over the course of many months, even when left unattended. This "land farming" of hydrocarbon wastes became widely practiced throughout the refining industry, although no particular notice was taken as to the mechanisms at work there. It was generally (and incorrectly) supposed that UV radiation from sunshine and atmospheric oxidation were the *sole* causative agents for the hydrocarbon destruction.

DISCOVERY OF HYDROCARBON-DEGRADING MICROBES

Inquisitive university and industrial microbiologists and other researchers investigating degradation of hydrocarbons in the natural environment found that certain microbes obtain their food and energy requirements from simple and complex hydrocarbon materials. Many of the common soil bacteria are among these hydrocarbon-degrading microbes.[4] Upon sampling and analysis, these were found to be the most active oxidizing organisms in typical land-farmed sludges. As more information was developed concerning land-farm management, it was determined that frequent tilling, watering, and occasional fertilizer applications drastically increased the rate of degradation. This emerging new approach to surface treatment of sludges and contaminated soils became known as *land treatment.* More recent research and study into this interesting remedial technology have resulted in the development of successful applications for cleanup of a wide range of hydrocarbons and organic contaminants. Some advances have been made in tailoring the same approach to the treatment of sites contaminated with various metals and inorganics.

In the case of groundwater, it was discovered that shallow aquifers, being in close contact and connection with the surface atmosphere, resisted long-term contamination and typically cleaned up at a faster rate than deeper aquifers, where oxygen diffusion is more limited. When polluted water was pumped from these deep aquifers to the surface and aerated, it cleaned up in short order. In addition, through subsurface introduction of oxygen and nutrients, these aquifers could be remediated in place.[5] This became known by the medicotechnical term *"in situ"* aquifer remediation (also sometimes referred to as the Raymond process, after its founder).

Early successes for bioremediation have come through treatment of crude petroleum, lube oils, simple and moderately complex organics, alcohols, fuels (gasoline, diesel, and fuel oils), some solvents, and nitrogen- and oxygen-substituted compounds. When these contaminants are present, bioremediation is often the most effective choice for consideration.

Comprehensive and responsive site evaluation, design, construction, and operation of any bioremediation project require the combined talents of many disciplines. Input from microbiologists, chemists, geologists, hydrologists, engineers, and experienced field remediation specialists is necessary for successful execution and closure of a "BIO" project. In fact, although biological treatment techniques can sometimes be utilized as the sole technology at some sites, it is often included in a remedial design as a pretreatment or polishing step to accompany other remedial techniques. These may include soil vapor extraction, air stripping, air sparging, and carbon adsorption.

Oil Spills

In this discussion, the subject of spills to surface waters is not directly addressed because it is the subject of other volumes. It is interesting, however, that petroleum and hydrocarbons released to surface waters can be as amenable to bioremediation as when the technology is applied to groundwater. Industrial wastewater treatment plants utilize biological treatment for the destruction of dissolved hydrocarbons (more concerning these applications to the present subject in Chapter 3).

Also witness the effectiveness of elevating the nutrient level at the water surface (or on beaches) of an ocean spill of crude oil. The Alaskan Oil Spill of 1989 was the site of a successful application of bioremediation technology[6] as was the Mega Borg Spill off the Texas coast in 1990.[7,8] As a matter of fact, treatment of spills in ocean and estuarine environments may proceed faster than those in fresh surface and groundwaters due to the increased natural populations of petroleum-degrading bacteria in the marine environment. The hydrocarbon-degrading fraction of the total heterotrophic microbial count in some ocean waters can reach 100%.[4]

In short, there is no shortage of hydrocarbon-degrading microbes in ocean or estuarine environments. Indeed, the marine environment possesses the richest indigenous population of these organisms, already acclimated and waiting for nutrients and oxygen sufficient to accomplish the work of remediation. It has been well established that the availability of nutrients is the limiting factor in microbial degradation of hydrocarbons in seawater.[10]

DISCOVERY OF METAL-DETOXIFYING MICROBES

Certain classes of microbes are known to alter the *redox potential* or valence state of some of the metals. Early understanding about the role of bacteria in enhancing the leaching of metal ores led to the discovery of microbial mining techniques such as *heap leaching*. Natural ores consist of metals that are fully or partially oxidized. Metals are extracted through methods that result in their reduction and/or solubilization. This was later discovered to take place as a function of microbial reduction and biosorption of metals. Bacteria do this as a natural consequence of their metabolism.

In contrast to organic pollutants, metal contaminants are not biodegradable. However, many important toxic metal contaminants can be transformed to less-soluble or even volatile species by bacteria. Today, the state of the art includes knowledge of the microbial ability to immobilize some metals and mobilize others, depending upon their physical behavior at various valence states. Bacteria can change the redox potential of metals under both aerobic and anaerobic conditions and can bioconcentrate them. Microbes employ several methods for reducing the toxic threat of metals, including (1) adsorption to living or dead cell walls, (2) ion-exchange at the cell wall, (3) extracellular precipitation, (4) forming organo-metallic complexes, and (5) using the metal as an electron acceptor during metabolism. An in-depth look at these processes and their potential for reducing the toxic risk of metallic pollutants is presented in Chapter 3.

THREE BIOREMEDIATION TREATMENT APPROACHES

Whenever bioremediation is contemplated as a cleanup technology, a serious consideration is choosing the particular mode of biotreatment that will offer the greatest advantage. Once the site evaluation has been completed and sufficient data have been gathered on the nature, location, and concentrations of site contaminants, a decision must be made on the specific treatment that will be employed for cleanup. There are three options for application of bioremediation. The choice between these application approaches will be a key influence on the success or failure of a project.

Generally, the indigenous microbes have the advantage for *in situ* soils and groundwater treatment, while introduced species often outperform native microbes in a bioreactor and in some land-treatment applications. This is not always the case, however, and careful considerations are in order when undertaking a bioremediation project to determine which approach is best for a given contaminant mix and set of site characteristics. Quite a bit of lost sleep and heartburn can be avoided during the project through the conduct of a proper site evaluation and a thorough bioassessment or treatability investigation (see Chapter 4).

Biostimulation

If the results of site characterization and biofeasibility testing have shown the presence of a viable native population or community of specific contaminant-degrading microbes already in the site, then biostimulation becomes an available option. It is the essence of this option that the native (indigenous) microbial community is found to be *able* to degrade the target site contaminants if managed properly. It is the object of the process design team to provide the environment and resources required by the native microbes to accomplish the destruction and/or toxicity reduction of the contaminants to acceptable regulatory levels. Stimulation of the native microbes for site remediation has resulted in the vast majority of bioremediation site closures in recent years.

Bioaugmentation

Whenever testing indicates that stimulation of the native community will probably ***NOT*** accomplish an acceptable treatment when applied to a specific site, a question arises as to whether the bioremediation design specialist can purchase and employ species of microbes that are artificially introduced into the soil and/or water.

During recent years, a tremendous amount of research effort has been expended to discover the mechanisms by which microbes degrade and metabolize hazardous organic chemicals. Whole volumes have been written describing the possible degradation pathways that come into play in the destruction of chemical compounds under many varying regimes of environment, substrate mixtures, and active consortia of organisms.[9,10] At least one text makes an exhaustive examination of the literature on bioremediation of solvents and hydrocarbons,[11] and another takes a look at pertinent field application case histories.[12] The result of these efforts has placed emphasis on the disparity that exists between degradation efficiencies observed under closely controlled laboratory conditions and those encountered in the field. Very often there is no realistic correlation between lab findings and the multitude of parameters that dictate the efficiency and rates of degradation that may be expected in a field bioremediation project. Furthermore, the results of many *in situ* remediations have indicated that contaminant degradation is more often accomplished by the native community of microbes acting in concert rather than a single "hero species" at work.

It is true that genetic engineering and artificial selection for microbes that are specific pollutant destructors are coming of age ***in the laboratory***. However, the effective *in situ* field application of these hero species has sometimes been poor to dismal. As more field experience is correlated with carefully controlled lab work, it becomes apparent that the reason for these limited successes is simply the fact that the native populations are already acclimated to the existing environment and have learned to utilize the pollutants present as food sources. Thus, they naturally have a competitive edge over introduced species that find the environment strange and available food resources unwanted. After addition, these introduced microbes often rapidly decrease in number and can become undetectable within a few days. However, this is not always the case and some sites have been quickly remediated through addition of specific degrading microbial products. In short, at the present time, the least that can be realized from their application is a few days' "jump start" on the job at hand, and some sites are good candidates for full-scale bioaugmentation. The key parameter that must be demonstrated when contemplating bioaugmentation is the ***viability*** of the proposed specific degrader organisms in the site environment.

Perhaps a time will come when further advances will see the availability of safe, adapted, or engineered organisms that have viability as well as capability to destroy specific pollutants in the field. That time may come sooner if proper funding and R&D programs intelligently address the present inadequacies in this important arena. Then, too, as the regulatory agencies implement more stringent registration and control requirements for added organisms (i.e., the pending TSCA regulations), the costs of development and postclosure monitoring may become too great a burden. Many developing treatment strategies have become unattainable due to such added costs as these. Regulators unfamiliar with the realities of microbiology have unjustifiable reservations (which have also been assumed by certain environmentalist groups) about the possibility that microbial manipulation might spawn "**the BLOB**" or some other etherial notion of "nature gone out of control." The fear of creating "**the BUG that Ate Pittsburgh**" must be controlled and sanely addressed if we are to see meaningful advances in bioaugmentation technology.

One of the battles ahead for those approaches that involve release of acclimated, genetically altered, or engineered microbes into the environment will be the forthcoming requirement of testing and registration of the organisms with the U.S. EPA. Proposed legislation under the Toxic Substances Control Act (TSCA) may require registration at considerable cost for development, controlled releases, and long-term monitoring after the cleanup project is terminated.[13]

Intrinsic Treatment

The third bioremediation option might be available when stimulation and augmentation have been found to be too expensive, too slow, or otherwise inappropriate. The results of a proper bioassessment will indicate when the first two options are not going to provide the required performance. In that case, the data might provide evidence that:

1. A capable microbial community exists at the site.
2. Required nutrients are available.
3. Requisite environmental conditions exist.
4. The contaminants will degrade naturally over time.

With these data in hand, an appeal to the regulatory authority might result in a permit to pursue the intrinsic option for this particular site. The approved remediation strategy will require regular monitoring for specific site contaminants, microbial health and numbers, and possibly other parameters as appropriate. Environmentally progressive states such as California have embarked on a new economically friendly strategy of considering intrinsic bioremediation whenever possible at sites that require expenditure of tax dollars for cleanup.[14]

RECENT ADVANCES IN BIOREMEDIATION TECHNOLOGY

As versatile as biological treatment may seem, it is sometimes difficult to apply to certain complex xenobiotic chemicals which are resistant to microbial degradation (i.e., highly halogenated compounds) and metals. At the writing of the first edition of this book, it had not been proven field-ready for most heavy metals. There have been, however, certain advances. Although the present text cannot examine this subject in detail, several notable examples presented in Chapter 3 show that the state of the art in bioremediation of metals, inorganics, and recalcitrant organic contaminants is rapidly advancing.

For instance, inorganic nitrates can be remediated biologically through application or encouragement of denitrifying bacteria to convert nitrates to nitrogen gas. In the process, some metals can be detoxified. Inorganic sulfates can be reduced to hydrogen sulfide by sulfate-reducing microbes, and at the same time, this bioconversion causes a highly reducing environment which can result in the detoxification of metal ions, and/or valence state alteration in toxic or soluble metals that render them immobile and biologically unavailable.[15] Several metals can be readily concentrated by algal cells. Pilot treatment systems have been designed and constructed for demonstrating this peculiar biotreatment technology.[16]

Successful demonstrations of effective and consistent field applications for bioremediation of metal-contaminated sites are providing accumulating evidence that the technique is both timely and economical. If development of the technique proceeds as hoped, the biological treatment of toxic metals could well be the most important breakthrough in environmental remediation in recent history. Several sites are undergoing treatment that anticipates regulatory closures in the near future.

THE CHAPTERS AHEAD

Since the authors intend to present the science and application of environmental bioremediation in terms of field-ready and proven concepts, the remainder of the book will address the "nuts and bolts" that need to be understood by those involved in design and operation of these systems. The technical discussions will be as brief and to the point as possible while giving fundamental concepts and information in the proper sequence they are needed to select, design, construct, operate, and monitor a typical bioremediation project.

Chapters 2 and 3 cover the basics of microbial nutrition, metabolism, degradation mechanisms, and common considerations for prediction of bioremedial end points. Chapter 4 outlines the essential elements in laboratory setup and desired testing capabilities for proper biotreatability assessments. Chapter 5 gives the step-by-step data requirements and essential design elements needed to complete a project. Chapters 6 through 11 are tailored to specific bioremedial applications and present the peculiarities of each design choice available. And, finally, Chapter 12 reviews relevant developments and probable future trends in bioremediation technology.

REFERENCES

1. *The Concise Columbia Encyclopedia*, 1983, Columbia University Press, New York, p. 907.
2. NETAC Bioremediation Panel, 1991, National Environmental Technology Assessment Corporation, Pittsburgh, PA.
3. Watkins, J.D., Secretary of Energy, 1991, Introduction to the Executive Summary, 1st Five-Year Plan for Fiscal '93–'97, (DOE Thirty-Year Plan).
4. Jones, J.G. and Edington, M.A., 1968, An ecological survey of hydrocarbon-oxidizing microorganisms, *J. Gen. Microbiol.,* 52, 381–390.
5. Raymond, R.L., Hudson, J.O., and Jamison, V.W., 1977, *Bacterial Growth in and Penetration of Consolidated and Unconsolidated Sands Containing Gasoline,* API Publication No. 4426, American Petroleum Institute, Washington, D.C.
6. Anon, 1990, Alaska's bill spill; Can the wilderness survive, *National Geographic,* June, Vol. 177, No. 1, pp. 5–43.
7. Anon, 1990, MEGA BORG oil spill off the Texas coast: An open water bioremediation test, *Combined General Land Office & Water Commission Report,* State of Texas, Austin, July, 13 pp. and Appendix.

8. Anon, 1990, "Combatting Oil Spills Along the Texas Coast: A Report on the Effects of Bioremediation," *General Land Office Report,* State of Texas, Austin, June, 14 pp. and three Appendices.
9. Rochkind, M.L., Blackburn, J.W., and Saylor, G.S., 1986, Microbial Decomposition of Chlorinated Aromatic Compounds, EPA/600/2-86/090, U.S. Environmental Protection Agency, Cincinnati, OH.
10. Gibson, D.T., ed., 1984, *Microbial Degradation of Organic Compounds*, Marcel Dekker, New York.
11. Norris, R.D. et al., 1993, *Handbook of Bioremediation*, Lewis Publishers/CRC Press, Boca Raton, FL.
12. Flathman, P.E., Jerger, D. and Exner, J.E., 1993, *Bioremediation: Field Experience*, Lewis Publishers/CRC Press, Boca Raton, FL.
13. Korwek, E.L., 1989, Federal regulation of hazardous waste treatment by genetically engineered or adapted microorganisms, in *BIOTREATMENT: The Use of Microorganisms in the Treatment of Hazardous Materials and Hazardous Wastes*, Hazardous Materials Control Research Institute (HMCRI), Washington, D.C., November 27–29, pp. 83–88.
14. **Anon.,** 1995, Recommendations to Improve the Cleanup Process for California's Leaking Underground Fuel Tanks (LUFTs), final report to California Water Resources Control Board, Lawrence Livermore National Laboratory, Livermore, CA.
15. Nuttall, H.E., Lutze, W., and Barton, L., 1996, Preliminary Screening Results for *In Situ* Bioremediation, International Conference on Bioremediation of Mixed Waste, Albuquerque Technical Vocational Institute, Albuquerque, NM.
16. Darnall, D.W. and Hyde, L.D., 1989, Removal of Heavy Metal Ions from Groundwaters Using an Algal Biomass, International Conference on Bioremediation of Mixed Waste, Albuquerque Technical Vocational Institute, Albuquerque, NM, pp. 41–46.

CHAPTER 2

Microbial Nutrition and Environmental Requirements

Now that the basics of treatment definition, history, and remedial approaches have been covered, it is time to introduce the major players and provide some basic information on just what microbes do in the soils, sludges, and groundwaters being treated that results in a successful bioremediation. It has been said that every biochemically synthesized organic compound is potentially biodegradable, and that some man-made compounds which are presently thought to be persistent (*recalcitrant*) may be found to be biodegradable through future discoveries. Also, while apparently unable to do so at present, microorganisms may develop the proper metabolic and enzymatic machinery to deal with newly made chemicals through *acclimation,* after having been exposed to them over sufficient time.[1] In principle, no organic compound has infinite persistence built into its chemical structure. In line with the same principle, somewhere there is a class of microorganisms that possesses the metabolic, enzymatic, or genetic potential to degrade every organic chemical.

MICROBIAL UBIQUITY

Outside of active volcanoes (and perhaps deep interstellar space), microbes are ubiquitous in every environment on earth. They are simply *everywhere.* Sterile natural environments on this planet are practically unknown. This availability of the *work force* for bioremediation makes virtually all sites potentially suitable from this standpoint. All sites contain microbes; the challenge is to entice those present (through *biostimulation)* into a beneficial mode of metabolism that will degrade the target contaminants.

If it is determined in the site assessment that insufficient numbers of specific degrader microbes are present at a specific site, then the addition of appropriate cultures (inoculation of the site with foreign microbes; *bioaugmentation)* might be entertained. A word of caution is in order here. The choice for bioaugmentation should be considered with the aid of sufficient laboratory data on site-specific degradation rates, product cost analysis, and monitoring requirements to justify the added expense. The authors have seen many vendors of these products come and go in the market place. The reality is that some commercially cultured organisms have been acclimated to contaminants that may not even occur at your site. As a consequence, some of the project failures that have resulted when organisms were added have been widely reported and *remembered,* while some of their successes (regrettably) were little noted and soon forgotten. It is to the credit of a few far-sighted technical business firms that the science of bioaugmentation has advanced during the last few years and now shows much improvement *for some applications.* The simple test of the matter is to be watchful for the sometimes wild and undocumented vendor claims as opposed to good quality, verifiable data on product viability, nutrient requirements, contaminant compatibility, survivability in the site environment, and specific degradation rates for site contaminants.[2]

Soils and groundwaters contain many kinds of microbes including fungi (molds and yeasts), protozoans, and bacteria. Of these common native (*indigenous*) microorganisms, it is the bacteria and fungi that account for the degradation of practically all of the hydrocarbon and organic contamination entering the natural environment. It is the systematic enhancement of this natural biological degradation of contaminants by bacteria that comprises the art and science of environmental bioremediation.

It has long been known that certain naturally occurring marine bacteria can quickly biodegrade petroleum entering the ocean environment.[3] Their presence accounts for the dramatic success in applying bioremediation to the oil-soaked beaches of Prince William Sound, AK, following the grounding of the *Exxon Valdez.*[4] More recent discoveries have revealed that many terrestrial soil bacteria are also quite efficient hydrocarbon degraders.[5] These comprise common soil and aquatic bacteria that also inhabit subsurface **vadose zone** soils and aquifers. In fact, viable bacteria, including many **aerobic** (oxygen-using) species, have been found at depths up to 1500 m below ground surface in deep soils and groundwater.[6] Virtually all groundwater and soil support a population of viable bacteria. Only in cases of extreme contamination by microbial toxins or solvents will a site be found sterile, and this is quite rare.

The presence of contaminant-degrading microbes in polluted soils and groundwaters is the vehicle for bioremediation. The manipulation of many other factors is involved in successful application. Key among these are the environmental parameters at a given site and the nutritional factors that will induce the biodegradation of target contaminants, and which will produce the desired remedial end point.

Here a word is in order concerning those who will insist that it is necessary to specifically identify the bacteria (down to genus and species) found in a site undergoing bioremediation. Outside of theses and dissertations, the wisdom of this exercise is doubtful. The reason is that so many of the naturally occurring bacteria are simply *unknown.* Bergey's *Manual of Systematic Bacteriology* contains a total of about 1.2 million known bacteria. It is estimated that there are no less than 5 million bacteria in existence, and probably many more. In other words, there is only about a 20% chance of being able to identify any specific bacterial organism, or, conversely, about an 80% probability of discovering an organism that is unknown. Since previously

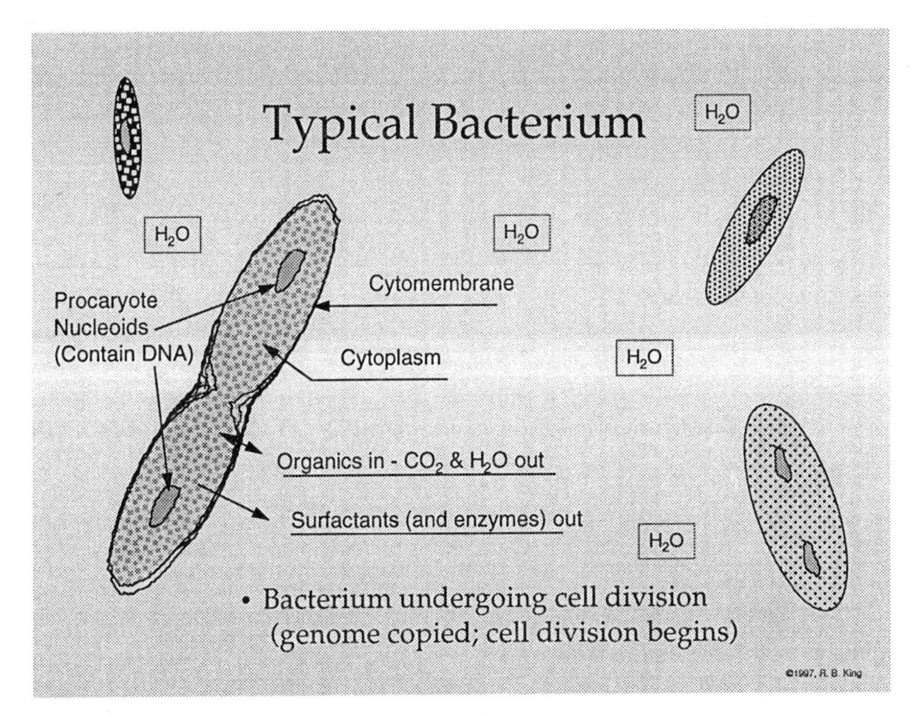

Figure 2-1 Typical bacterium

unknown organisms cannot be specifically identified, this practice is seldom productive. The expense of the exercise may not be too great in all cases, and the information may even be helpful for some sites, but not always. At any rate, it is optional at best.

MICROBIAL NUTRITION

It might be useful to think of each bacterial cell as a small chemical factory in which raw materials (the target contaminant compounds or chemicals in a contemplated remediation site) are converted to harmless, less mobile, less soluble, or less toxic end products. In many cases, these contaminants can supply all the food requirements bacteria need for energy and growth (Figure 2-1).

All life forms require food, water, and a suitable environment in which to live, grow, and multiply. Food for microbes must provide a source of carbon for building essential biochemicals and cellular components (growth), and which is also a source of energy. The carbon source (food) for our microbes is also termed the *nutritional substrate* and should be restricted to the contaminant; i.e., no other carbon source should be made available whenever possible. The addition of supplemental organic materials to a site, for whatever reason, must be evaluated in the *biotreatability study* to ensure the microbes do not preferentially degrade the added carbon source rather than attacking the target contaminants.

In general, microbes will first attack and degrade compounds that offer the greatest amount of energy at the least expenditure of biochemical effort on the part of the microbe. How do they know to do this? They know it chemically (due to thermodynamics), because the laws of thermodynamics are built into every molecule in the universe. This leads to a situation in which a light hydrocarbon (i.e., a C_7 compound) will biodegrade far sooner and faster than a heavy end compound (say, a C_{20}). So the *light ends* and 0- or N-substituted compounds always go first, followed by the *aliphatics* and multiring (PAH) compounds. The last to go are the long-chain (*paraffinic*) hydrocarbons.

Microbes are categorized by the types of substrate they can use; i.e., since bacteria are not photosynthetic and cannot utilize sunlight for energy, they must depend on chemical substrates. The substrate may consist of organic carbon of many varieties (including hydrocarbons), inorganic carbonate, or CO_2.[7] Microbes which utilize only organic carbon substrates are termed *heterotrophic* and these comprise the great majority of bacteria. Those which can utilize inorganic carbon substrates are termed *lithotrophic* and can metabolize and grow in environments which contain little or no organic carbon. The bacteria that are most useful in bioremediation of organic contaminants are the hydrocarbon-degrading heterotrophs. Also useful are the lithotrophs that can reduce metal solubility or toxicity and those that can perform reduction of nitrates and sulfates.

MOISTURE

Moisture is essential for growth and multiplication of microbes. Lacking mouths, microbes are limited to soluble materials that can be transported across their cell membranes into the interior cellular fluid where digestion takes place (although digestion can take place outside bacterial cells in certain instances; *extracellular digestion).* Optimal growth rates take place where there is sufficient moisture in liquid form to solubilize the substrate for transport. However, some bacteria and fungi, which are single-celled plants, can utilize the moisture present in air of high humidity for maintenance of cellular functions. Generally, when treating soils in the unsaturated (*vadose*) zone, moisture content is optimal between 10 and 25% by weight (20 to 50% field capacity). This provides sufficient liquid moisture within the soil pore spaces for optimum cellular metabolism, reproduction, and growth. However, lower moisture content in arid soils and clays are still able to support active populations of contaminant-degrading microbes.

A note of caution is due here. Soil moisture can be quantified in several ways. Generally, 10 to 25 wt% is optimal for aerobic bioremediation of soils. This translates to a typical value of between 20 and about 70% of the moisture *holding capacity* of a soil (depending on grain size). In gardening circles, this is referred to as *friable* soil. In geology, the parameter is given as percent of pore space filled with water (percent saturation). In other words, when all the pore space has been filled, holding capacity is at maximum (100% saturated). In soil bioremediation, soil moisture *must* be held within these limits (10 to 20% wt.) for effective *aerobic* biodegradation. Overwatering can saturate the soils and cause *anaerobic* conditions by driving out the air. This event will stop any aerobic degradation in progress and stall the remediation. In this case, the site must be returned to proper moisture levels before aerobic biodegradation can be re-established.

ENERGY FOR METABOLISM

Given sufficient moisture and a suitable substrate, microbes further require a source of energy to propel metabolic reactions. *Autotrophic* organisms such as green plants and algae, being photosynthetic, require sunlight energy. They will not metabolize in the dark and, therefore, are of no assistance for remediating subsurface soils and groundwater. *Protozoans* (single-celled animals) which graze on other microbes and require high amounts of free water and oxygen, are also not expected to be of any assistance in degradation of contaminants in the subsurface.

That leaves the bacteria and fungi as the microbes of importance for bioremediation. Without sunlight, metabolism in darkness requires chemical energy derived wholly from oxidation-reduction reactions. In other words, because they cannot utilize sunlight, they must obtain their energy from the organic chemicals that were made when phototrophs originally captured energy from the sun for carbon fixation. Thus, the *heterotrophic* bacteria and fungi are capable of the type of metabolism that is able to biodegrade hydrocarbons, organic chemicals, and complex organic substrates. These heterotrophic microbes derive metabolic energy from many kinds of organic substrates and are extremely adaptable to hostile environments.

THE ROLE OF OXYGEN

Heterotrophic bacteria obtain their energy requirements from the available carbon source substrates through electron transport pathways. Whenever a microbe degrades an organic substrate for its carbon content and energy needs, electrons are removed from the substrate and passed on to a suitable electron acceptor. This simply means that reduced carbon atoms are oxidized (releasing energy) by giving up electrons through the action of biodegradation by microbes. This electron transfer amounts to the transfer of hydrogen atoms to a suitable electron acceptor which is then chemically reduced. Whenever these electrons are passed directly to molecular oxygen, the degradation pathway is termed *aerobic.* In this case, oxygen is the terminal electron acceptor. The reaction creates both carbon dioxide (carbon is oxidized) and water (oxygen is reduced).

In aerobic metabolism, oxygen is the terminal electron (hydrogen) acceptor. This requires availability of some form of free oxygen and the microbes utilizing this degradation pathway are termed *aerobes. Microaerophiles,* though aerobic, can tolerate only low concentrations of oxygen. When something other than oxygen is the terminal electron acceptor, the reaction is termed *anaerobic.* In general, aerobic degradation proceeds at a faster rate than does anaerobic degradation.

When free oxygen is not available, many microbes can utilize other means of oxidizing carbon for energy and growth through the use of various anaerobic pathways (i.e., those that can utilize the oxygen bound to nitrogen in nitrates for the final electron acceptor). These microbes are termed *facultative anaerobes.* These organisms do not need or use O_2, but sometimes grow better in its presence.

So, then, aerobic organisms depend on oxidation of suitable substrates where O_2 is the final electron acceptor, and facultative anaerobes can utilize the oxygen combined in nitrates. Some organisms, though anaerobic, can tolerate oxygen and grow equally well whether it is present or not. These are termed *aerotolerant anaerobes.* On the other hand, strict or *obligate anaerobes* depend on oxidation of substrates where the final electron acceptor is something other than free oxygen or nitrate and can tolerate no oxygen at all.

Fermentative organisms carry out oxidation-reduction reactions between organic electron donors and acceptors. Oxygen may or may not participate. All of the foregoing metabolism types will be covered in greater detail in Chapter 3.

All of these reactions and metabolic pathways (transfer of electrons from food substrates to electron acceptors) are mediated and accomplished through the action of enzymes. These constitute the magic in microbial metabolism that converts hazardous organics to harmless carbon dioxide and water and can change the redox potential of toxic metals. So enzymes act as chemical catalysts in the oxidation and reduction reactions that constitute microbial metabolism. Resident or introduced microbes must possess the necessary enzymatic software in their genetic material (DNA) in order to produce the enzymes that can degrade the site contaminants or change the chemical valence state of metal atoms.

Whenever an organic substrate is biodegraded, microbes obtain energy and the carbon metabolite biochemical building blocks necessary for growth and reproduction. In the calculation of mass balance for carbon, that portion of the original carbon of the target organic substrate utilized for nutrition ends up as more cellular biomass (about 65%). The portion which supplies energy requirements for cellular metabolism is converted to carbon dioxide (about 35%).

Major nutritional factors that are essential compounds of importance for microbial nutrition in bioremediation are

Nitrogen (as ammonia)
Phosphorus (as orthophosphate)

It is here that inexperienced personnel involved in the application of field bioremediation go astray. The authors are constantly contacted by clients who are the owners of a site undergoing cleanup by a company claiming to be familiar with bioremediation. It is amazing how many of these sites are simply out of nutrients, because the operators are not monitoring or supplying the correct forms of nitrogen and phosphorus. As a general rule, nitrogen must be present in the form of ammonia, although certain bacteria can utilize nitrate or nitrogen gas. The simple fact is that bacteria are quite able to utilize ammonia for direct incorporation into amino acids etc., while nitrate must first be reduced to ammonia for use, or nitrogen gas must be fixed by reduction to ammonia. These reactions converting nitrate and nitrogen to ammonia are a waste of energy for the microbe and will slow the process of bioremediation. Ammonia nitrogen is the preferential nutrient. When it is lacking, the remediation can slow to a halt.

Usually, orthophosphate (phosphoric acid) must be available as the source of phosphorus. Analysis for total phosphorus will often give high numbers while the available orthophosphate may have been depleted. In this context, either N or P may become the limiting factor in sustaining bioremediation.

Essential minor elements are those needed in trace amounts:

Sulfur
Potassium
Magnesium
Calcium
Manganese
Iron
Cobalt
Copper
Molybdenum
Nickel
Zinc

It is the macronutrients (N and P) that are of major concern in the field application of bioremediation, because the minor and trace elements are usually present in sufficient amounts in the natural environment and are almost never the limiting factors in the field. Therefore, the assumption can usually be made that the limiting nutrients will be ammonia nitrogen and orthophosphate. It is useful to portray the relationships between *microorganisms*, their required *nutrients*, and an available carbon source (*food*) in terms of the Bioremediation Triangle, illustrated in Figure 2-2. By substituting specific names for these components (see Figure 2-3) we can arrive at a graphic understanding of the necessary ingredients for a successful application of this technology. It is the lack of a basic understanding of these simple relationships that accounts for most of the failures we see in field attempts at bioremediation.

ENVIRONMENTAL REQUIREMENTS

In addition to strict nutritional requirements, microbial growth and metabolism are keenly affected by the chemical and physical environment. The microbial kingdom is extremely varied and versatile in its response to environment and, within limits, some groups can survive and grow even in extremely harsh conditions.

The physical environment offers many challenges to bacteria. For instance, *Thiobacillus* species active in sulfur oxidation grow well in concentrated sulfuric acid (pH <1). On the other extreme, *Bacillus* species active in metabolism of urea prefer pH in the range of 9 to 11.5 and cannot live at less than pH 8.5. Cultures of *Nitrobacter* and *Nitrosomonas* multiply and grow at pH 13. Most microbes, however, prefer pH 6 to 8 for maximum rate of growth.

Hydrostatic pressures of many hundreds of atmospheres are present in the deep subsurface and in deep ocean sediments supporting prolific bacterial growth. Certain thermophilic bacteria can metabolize and proliferate in the deep ocean at temperatures up to 115°C. Many bacteria can be frozen and kept at temperatures as low as −194°C for many years without great loss of viability upon thawing. Although

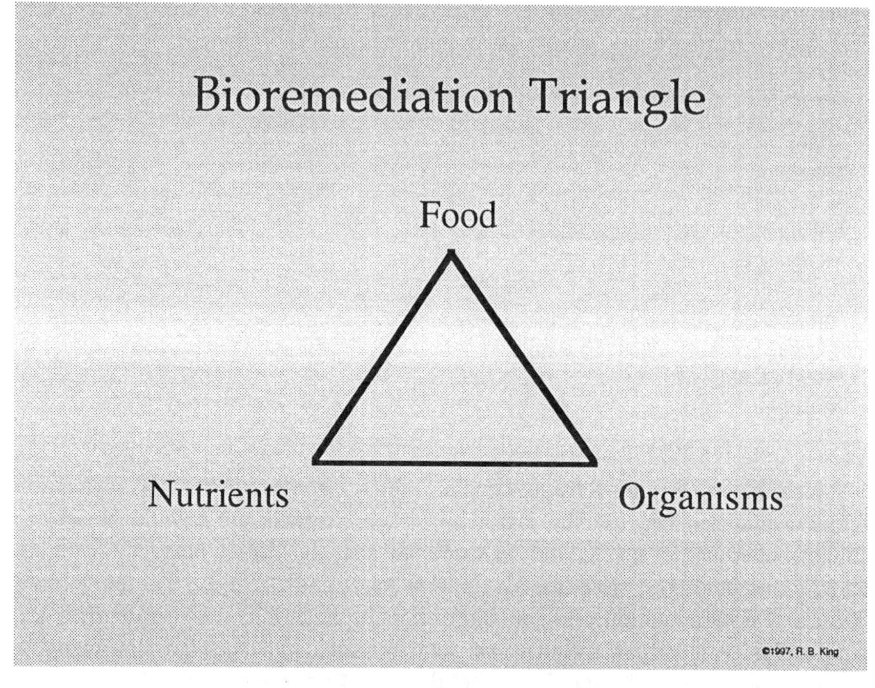

Figure 2-2 Bioremediation triangle

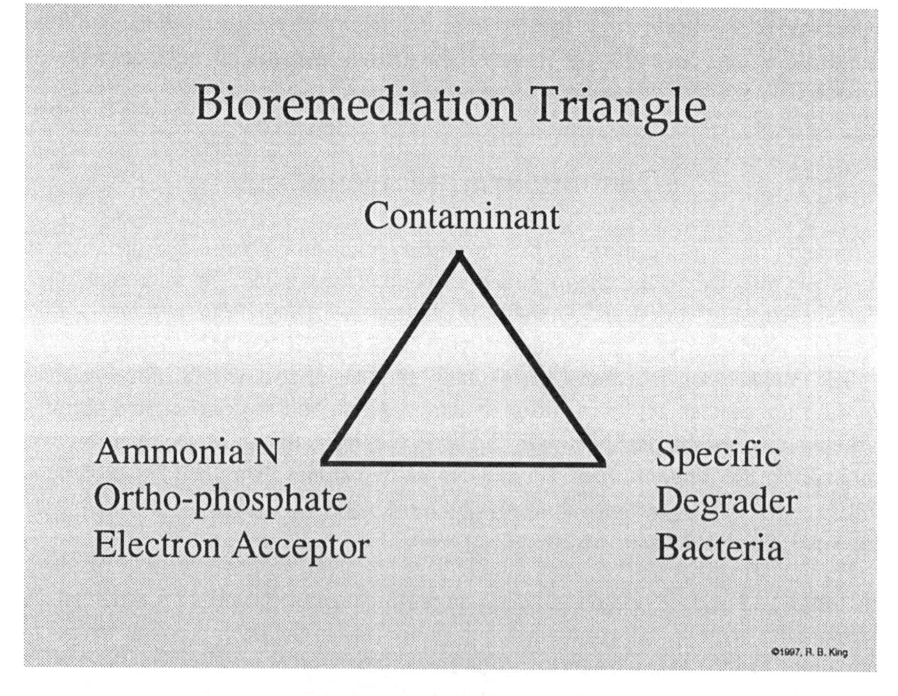

Figure 2-3 Bioremediation triangle

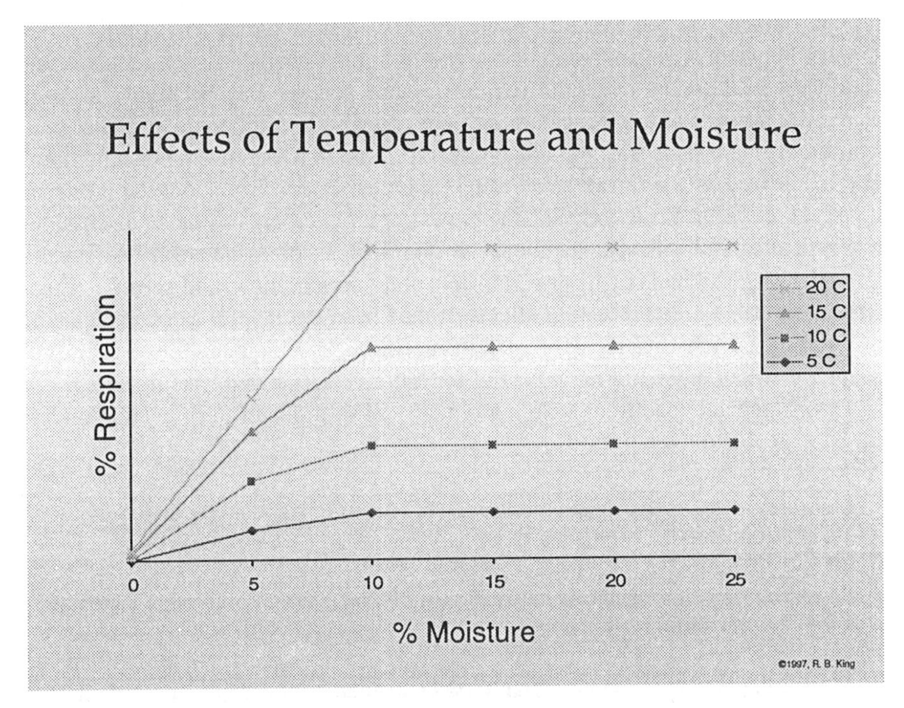

Figure 2-4 Effects of temperature and moisture

dormant with no active growth and metabolism, desiccation to dryness under low heat will also preserve bacterial suspensions. Notwithstanding these extremes, most microbes require temperatures between 0 and 80°C for metabolism with an optimum between 20 and 50°C. One of the authors worked with a native Alaskan soil bacterial community that was quite at home degrading petroleum at temperatures just above freezing in a bioreactor application. When the operating temperature was raised to "textbook levels" the degradation stalled. It is unlikely that lithostatic pressure will ever be of concern in bioremediation. However, the ambient temperature within a site matrix can dictate effective rates of biodegradation. Generally, as temperatures fall below optimum, the metabolism rate will fall below a satisfactory level. The relationship between temperature and moisture on microbial respiration rate is depicted in Figure 2-4.

Dissolved solutes determine the osmotic pressure affecting the microbial environment. Most common bacteria can withstand wide ranges in osmotic pressures. For example, the aerobic soil bacteria *Aerobacter aerogenes* can grow in media with salt concentrations ranging from less than 0.1% to about 12%. All marine bacteria require about 3.5% salt and cannot grow at less than 1% salt. Organisms found growing in strong saline environments, termed "halophilic" (salt loving), can grow in saturated (30 to 35%) solutions. An underground microbiological survey of the Waste Isolation Pilot Plant in New Mexico (the Department of Energy facility designed for placement of some low-level radioactive wastes in underground salt formations) revealed that at least 48 species of aerobic bacteria are well established in the rock salt.[8,9]

The influence of the chemical environment is just as varied. Physiological differences between organisms account for vast differences in their response to specific chemicals. A chemical that is a nutrient for one species may be highly toxic to another. A nutrient at low concentration may likewise become toxic at higher concentrations in the same species, and, conversely, a toxin found in low concentrations may be utilized as a substrate. One cannot therefore generalize about the influence of a specific chemical substance on the growth of all microorganisms. In discussions of physical and chemical constituents, it therefore becomes necessary to specify particular physical and chemical parameters and concentrations and specific organisms in describing or forecasting effects. It thus becomes helpful to obtain specific information on the toxicity of contaminants through a biotreatability study (see Chapter 4) when assessing the potential for bioremediation at a specific site.

Environmental conditions within a typical soil or groundwater remediation site can be expected to be

Moisture — Dry to completely saturated (0 to 35 wt%)
Temp — 0 to 43°C
pH — 5 to 9
Eh — Initially oxidizing (air saturated) in soils; (few aquifers are ever truly anaerobic)
D0 — in groundwater 2 to 6 ppm.

Moisture, pH, and redox potential (oxygen available for aerobic metabolism) are control parameters of concern in the design and execution of a bioremediation project. Ambient temperatures must be known or anticipated in order to predict the rates of degradation, as this is the least controllable parameter.

Generally speaking, as environmental conditions change, so does the potential for active metabolism by specific groups of microbes; e.g., as changes occur in pH, temperature, redox potential, oxygen availability, etc., certain groups become favored and certain others slow or cease metabolism. Thus, there are natural fluctuations in organism diversity and population numbers as a function of time in virtually any microbial community unless conditions are strictly controlled. It is the strict control of conditions (along with proper analytical monitoring) that becomes the key issue in performing a successful bioremediation in the field. The conduct of a biotreatability study, when needed, will define the present state of a site and predict the probable success of treatment (see Chapter 4).

* * *

Let's pause a moment here. So far we have seen that microbes exist in virtually all natural environments, possess specific physical, chemical, and nutritional requirements, and are very versatile and adaptive to hostile environments and changing nutritional sources. From an operational viewpoint, these are the major considerations for field control and should provide a basic understanding of bioremediation. The sections that follow are given as a deeper discussion of microbial dynamics and (1) demonstrate the growth and maintenance of a healthy microbial population of contaminant degraders; (2) show that microbes acclimate to new site conditions and utilize new food sources; (3) demonstrate microbial persistence over long time periods without food or water; and (4) show that microbes are well adapted to ionizing radiation and are able to degrade the organic portion of radioactive mixed waste. Chapter 3 will provide the details of the various forms of microbial metabolism useful in bioremediation and detoxification of contaminated sites.

* * *

MICROBIAL GROWTH

Since bacteria are commonly found in all soil and groundwater, and it is well known that many of them can degrade chemicals of environmental importance when properly stimulated, it first becomes important to understand that microbes can quickly adjust to alternate modes of metabolism that may be required by changes in the prevailing conditions of environment and available carbon sources. This ability of microbes to "switch gears" to adapt to new conditions is termed *acclimation* and accounts for the wide variety of environments in which they are found and for the many carbon sources they can utilize for food.

Acclimation

Microbes are sensitive to abrupt changes in their environment. Whenever the physical or chemical environment is suddenly changed, there is a transition time during which the microbial population gets reacquainted with the new conditions. The growth and metabolism of the organisms may be temporarily interrupted. This *lag phase*, or "acclimation" period, is caused by the microbial population shifting its metabolism (gene induction or derepression, among other things) to compensate for the altered environment.

For instance, when a spill of hydrocarbon fuel or solvent occurs into the soil or groundwater, the resident microbial population is presented with a newly arrived chemical in its environment that it had not "seen" before. As all hydrocarbons and organics have potential as a substrate (food source) for bacteria, some of the microbes in residence may make internal enzymatic changes (like gene expression) in order to deal with the chemical newcomer. These enzymatic changes may take place for two reasons: either the population attempts to transform the organic compound (perceived as a toxic threat in its present form) into an innocuous form, or the microbes may attempt to degrade (mineralize) the compound for energy and growth.

Generally speaking, the longer a contaminant has been in contact with a microbial population, the better acclimated will be that population. In short, sites with a long history of contamination have the better likelihood of possessing strongly acclimated contaminant-degrading native microbial populations (see Figure 2-5).

Logarithmic Growth

Once the population of specific degraders has acclimated and sufficient nutrients and all other growth factors have been provided, they typically begin what amounts to a logarithmic rate of numerical increase. During this *log growth phase,* the microbes will undergo a doubling in number per unit time. The slope of this curve

Microbial Growth Dynamics
for a Batch Culture

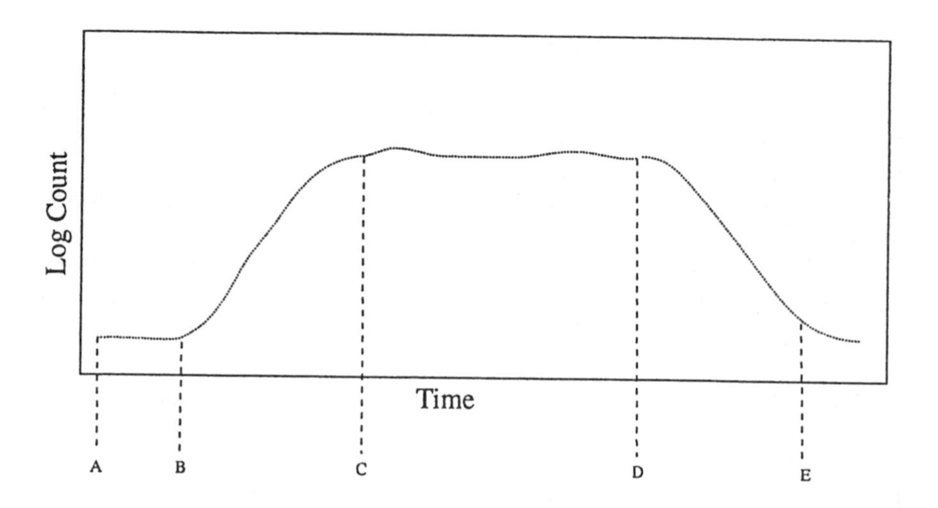

A - Lag phase begins; microbes begin acclimating to new environment

B - Log Growth phase begins; acclimation complete

C - Stationary phase begins; cell division rate equals death rate

D - Death phase begins as cell division slows; food source depleting

E - Cell count returns to normal; low nutrient conditions

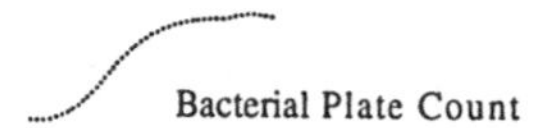

Figure 2-5 Microbial growth dynamics for a batch culture

(the growth rate) will be dependent on many factors — too many to list. The ultimate concern for the remediation specialist is to maximize this growth rate by doing whatever is necessary (within the law) to accommodate the microbes. The initial control parameters will emerge from the biotreatability testing.

Fine tuning will come as a result of monitoring the project and making any midcourse adjustments that seem to be appropriate. For the project to be successful microbial metabolism in this early stage of the remediation must result in the establishment of a steady rate of decline for the contaminants. The mere observance of an increase in microbial numbers does not mean the site is cleaning up. The microbial population might instead be using some other substrate and leaving the contaminants intact. Therefore, it becomes necessary to monitor both the microbial counts and the contaminant concentrations. These are the parameters of interest to the regulators, and they constitute the data that can substantiate that bioremediation is indeed taking place. By the same token, data indicating decline in contaminant concentrations do not, in themselves, indicate successful bioremediation. One of the authors once sat before a group of engineers who were running a project at a tie-treating plant for bioremediation of creosote. The engineers presented their site data, and in the end (as no microbiologist was included in the project team, or even consulted) the only data presented that suggested bioremediation had taken place were figures for nutrient additions and curves showing slowly falling contaminant concentration. These people were doing "bioremediation by assumption." They assumed (incorrectly) that because they had added the textbook quantities of proper nutrients and observed pollutant concentration to decline, they were doing bioremediation. The monitoring data simply did not substantiate that anything biological was happening at all. It should be obvious to anyone who has read this far that monitoring data must include biological information as well as physical and chemical measurements in order to tell what might be going on at a site. The regulators will insist on good test data to verify that the contaminants were degraded, and not just displaced or moved.

Stationary Phase

Microbes that are degrading the contaminants will continue to increase in number until they find a balance in terms of available food, nutrients, habitat, and the concentration of their own waste products, which can become quite toxic if allowed to accumulate. Fortunately, it is often the case that other microbes in the site soil or water matrix can degrade or detoxify these wastes. The point at which all these considerations reach equilibrium will determine where the microbe population count stabilizes. When this point is reached, the population numbers will remain stationary over time (cell division and death rates are equal) as they degrade the target contaminants (see Figure 2-6). This is the condition during which the contaminant degradation rate is at its maximum. All site activity is centered around maintaining this status quo to minimize the time to completion (several nutrient additions may be required). All the while, however, environmental conditions are steadily worsening for the degrader microbes.

Death Phase

As the target contaminants diminish to low levels, waste products begin to accumulate and the microbes begin to experience local overcrowding. The competition for available resources gets severe and some of the microbes just cannot keep up. The death rate exceeds the cell division rate and total microbial counts begin to decline. At this point, the target contaminants *should have been* degraded to below action level concentrations and the microbe population should return to numbers that were observed prior to treatment. When plate counts indicate a decline in the microbial population, further reductions in the contaminant will take place slowly, but they do not necessarily stop. Any remaining pollutants may very well remediate to lower levels over time and continued monitoring will substantiate this.

Microbial Growth Dynamics
and Contaminant Destruction
in an Ideal Biodegradation

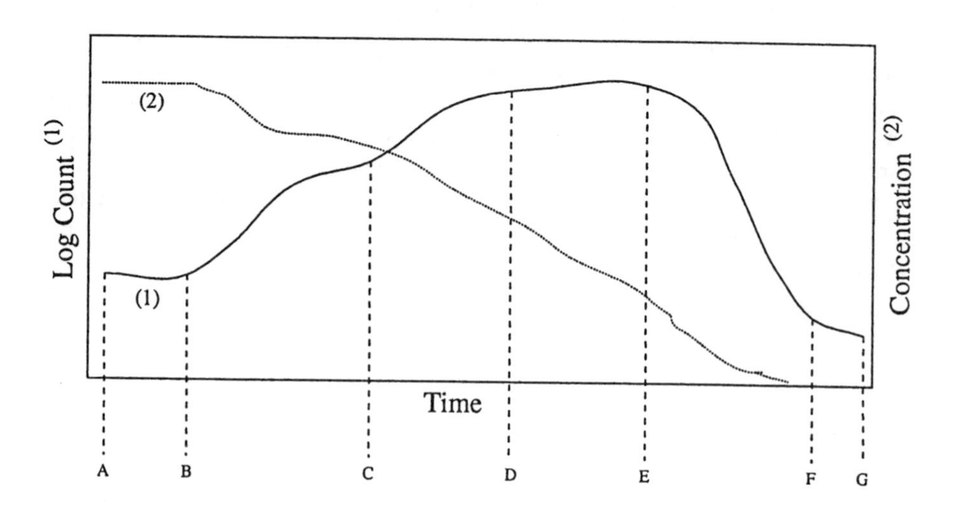

A - Contaminant introduction (spill occurs, etc.) ; lag phase begins

B - Acclimation complete, nutrients available, log growth & degradation begin

C - Second nutrient addition, continued growth & degradation

D - Growth stabilized, stationary phase, active and brisk remediation

E - Food source depleting, bacterial count declines as death phase begins

F - Remediation complete, biomass degenerates, population stabilizes

G - Endpoint of project with clean site

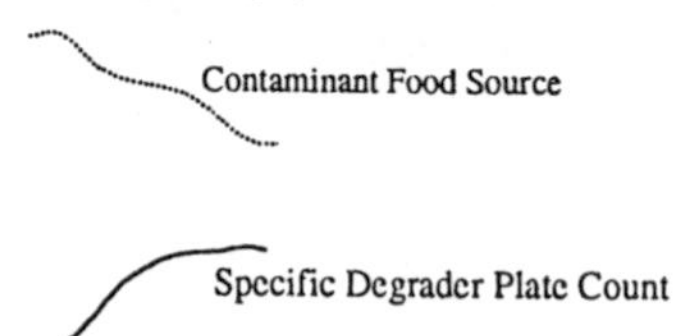

Figure 2-6 Microbial growth dynamics and contaminant destruction in an ideal biodegradation

BACTERIAL PATHOGENS

One thing must be made perfectly clear. Bioremediation does not employ known disease-causing (pathogenic) bacteria for the cleanup of contaminated sites — *ever!* The reader must understand that many materials, when misused, can become toxic or deadly. Common table salt is lethal in large-enough dosage, and pure water or air injected into the bloodstream can cause instantaneous death. Water, salt, and air are not commonly considered to be toxins or deadly, but situations can arise in which they become so. Likewise, virtually any microbe can kill a host if improperly administered.

One of the authors once read an article in a soils magazine in which its writer, a vendor of bioremediation bacteria, warned potential users of this technology to beware of the possibility that bioremediation, improperly applied by others, could cause potential danger through cultivation of human pathogens which might threaten public health. Obviously, he considered that his product was not in the dangerous category!

In the first place, common pathogenic microbes are of no use in bioremediation. Second, the presence of virulent pathogens at a particular site is the result of prior human or animal vector contact and is not initiated through bioremediation practice. Third, pathogens that may reside in soils and groundwater at a particular site are not stimulated through application of bioremediation technology, because their basic nutritional requirements cannot be provided there. Fourth, as the reading of any introductory text on microbiology will reveal, they require a human or animal vector internal physical environment near 99°F. in order to germinate, reproduce, and cause infection. Fifth, the soil and groundwater environments (being completely foreign to the requirements of pathogens) will simply not support their biochemical needs. And last, the practice of common prophylaxis and perhaps the employment of personal protective equipment (PPE) under the OSHA regulations required for associated site work will prevent infection of workers.

It might also be pointed out that *farming* is a common practice that entails the intimate contact of humans with the soil and groundwater. Farming also utilizes the excrement of domestic animals for soil conditioning and stimulates the growth of soil microbes for nutrient transport and cycling. Precautions against pathogens do not constitute a high priority by those working in agriculture.

MICROBIAL SURVIVAL

In order for any attempted bioremediation project to result in success, the bacterial population being employed must remain alive and multiply in the site matrix. Survival, or viability, of the workforce is simply mandatory. Obviously, many parameters are of importance to microbes and we have only mentioned the very most important of these. Simply stated, biotreatability testing will show whether the desired result can be obtained (microbe plate counts will indicate whether or not success can be anticipated). In many sites, testing will show that the native population of bacteria can do the job if stimulated to greater numbers and sufficient appetite.

Microbiologists have long known that counting colonies on agar plates is not a reliable method of estimating the *total number* of microbes in a sample. There are simply too many variables that limit this method. It can be clearly seen from microscopic analysis that there are many bacteria that just do not grow under a given set of circumstances. Sometimes only a small percentage of microbes present in a sample will form colonies; the others appear to be in some form of dormant state.

Those organisms that are visible can sometimes be stimulated to grow colonies and sometimes they can't. So, for our purposes here, those bacteria that respond and grow under site conditions are useful or viable, and those that do not grow are worthless. Thus, the claim for great numbers of microbes in a commercial product in itself is meaningless. The important information comes from viability testing of specific degraders in the site-specific soil or groundwater.

Many microbes can survive long periods of harsh environment through the mechanism of endospore formation. Whenever stress pressure is sufficient on some microbes (as occurs in periods of drought, starvation, etc.), metabolism rates fall to a point where sporulation is triggered. This causes the microbe to expel excess water and cell contents, concentrate genetic material, and form an encapsulated envelope which is impervious to the outside conditions. In this spore form, the organism is dormant and does not metabolize, respire, or reproduce. It simply rests and waits for favorable environmental conditions for germination. Spores can withstand considerable trauma over long periods of time including heating, desiccation, freezing, toxins, and high salt environments.

Furthermore, organisms under conditions of constant stress seem to select for even more resistant strains. When desiccated to dryness, spores remain viable for centuries, and possibly millennia.[10] In addition, they can become much more resistant to radiation and toxic chemicals. It is due to this ability to form spores and resting stages that microbes can survive severe conditions and this accounts for their presence in virtually any natural setting. It is rare indeed to find any environment, however polluted, that does not contain viable microbes.

* * *

At the close of this discussion of how microbes interact in the environment for our benefit, it is necessary to clarify and put in perspective one of the common colloquialisms in use for describing beneficial microbes. The term *"bugs"* is sometimes given (by the uninitiated) to certain organisms of use in bioremediation. The authors have been within earshot of mumbled conversations in which this oft-used term was employed to describe the bioremediation work force. For a time, it seemed an inappropriate colloquialism and quite offensive, as it came closer to a description of, possibly, an insect. At length, however, some intuitive author or speaker was able to put the term into perspective. It seems to be derived as an acronym for **B**acteria **U**nder **G**uidance and **S**upervision. This having been finally clarified, the authors wish to authenticate its accepted occasional use in the street vernacular, **BUT NOT IN THE WRITTEN WORD,** ***OR IN THIS BOOK*****!**

REFERENCES

1. Dagley, S., 1984, *Microbial Degradation of Organic Compounds*, Chapter 1, Marcel Dekker, New York.
2. Norris et al., 1994, *Handbook of Bioremediation*, Section 11, Lewis Publishers/CRC Press, Boca Raton, FL.

3. Mulkins-Phillips, G.C. and Stewart, J.E., 1974, Distribution of hydrocarbon-utilizing bacteria in northwestern Atlantic waters, *Can. J. Microbiol.*, 20, 955–962.
4. **Anon.,** 1990, Alaska's big spill; can the wilderness survive, *National Geographic*, June, 177(1), 5–43.
5. Jones, J.G. and Edington, M.A., 1968, An ecological survey of hydrocarbon-oxidizing microorganisms, *J. Gen. Microbiol.*, 52, 381–390.
6. ZoBell, C.E., 1946, Action of microorganisms on hydrocarbons, *Bacteriol. Rev.,* 10, 1–49.
7. Prescott, L.M., Harley, J.P., and Klein, D.A., 1993, *Microbiology,* 2nd ed., Wm. C. Brown, Dubuque, IA.
8. King, R.B., 1988, Microbial Degradation of TRU Mixed Waste in the Waste Isolation Pilot Plant, Internal Report for Westinghouse Electric Corp., Carlsbad, NM.
9. King, R.B., 1989, Analysis E; Effects of Microbial Degradation of Organics in the Exploratory Shaft, Yucca Mountain Mined Nuclear Waste System, Nevada, Internal Report for Sandia National Laboratory, Albuquerque, NM.
10. *Bergey's Manual of Bacteriology,* 1989, p. 1105.

CHAPTER 3

Microbial Destruction of Environmental Pollutants

Now that the reader is familiar with some of the major responses of microbes to their physical and chemical environment and their nutritional requirements, a clear understanding of the mechanics of just how microbes degrade or detoxify contaminants and reach acceptable reductions in pollutant concentrations during a bioremediation project is in order. Without a knowledge of these mechanisms, it will be difficult or impossible to apply bioremediation technology with any consistent success. An exhaustive discussion of the many degradative pathways known to exist for organic compounds and solvents can be found in other detailed texts.[1-3] Microbial detoxification of metals (changes in oxidation state), although well documented and understood, is a lesser known application technology for field bioremediation.[4]

BIOMINERALIZATION VS. BIOTRANSFORMATION

In the majority of bioremedial applications, and certainly in all site remediations, the object of treatment should always be *complete destruction* of the contaminants, if possible. This, of course, applies to organic contamination because metals cannot be destroyed, but only changed chemically (and sometimes biochemically) in their valence states. Therefore, bioremediation of metals always implies a *transformation* rather than destruction. Another term for complete destruction is *mineralization*, or reducing the contaminant to its basic mineral constituents. In the case of hydrocarbons, the end point reaction products of biological mineralization will be carbon dioxide and water.

If the biodegradation of an organic contaminant goes only part way to mineral (obtaining only a partial degradation), then the contaminant is said to have been *transformed*. This would be the case if the end point attained resulted in formation of some other less complex chemical. For instance, if phenol is degraded to benzene and no further, or if trichloroethylene (TCE) goes to vinyl chloride, these biotransformations could result in worse site conditions than existed before the remediation began. Even though phenol and TCE are toxic, benzene and vinyl chloride are both toxic *and* carcinogenic.

The point is that caution should be the watchword whenever the term "biotransformation" is seen on advertising literature or in presentations relating to the end point to be obtained in a field bioremediation for organic contaminants. If only a partial biotransformation of an organic contaminant is obtained, then site closure (with certain exceptions) might be doubtful. A proper biotreatability investigation should be pursued in order to identify potential problems.

BIODEGRADATION PROCESSES FOR ORGANIC COMPOUNDS

The soil and groundwater bacteria, being nonphotosynthetic microorganisms, are dependant on energy-yielding oxidation-reduction reactions for their source of energy. The amount of energy available to the microbe depends on the nature of the carbon source and the metabolic pathway utilized by the organism. It is a reality of biophysical thermodynamics that the "order of battle" for microbial degradation of organic contaminants will favor the lower molecular weight compounds. Therefore, in a site containing a contaminant mix, the *light ends* will be degraded preferentially before the *heavy ends* are metabolized. The biochemical energy budget favors easily degraded compounds and saves the more difficult substrates for hard times. Microbial enzyme systems are controlled by a genetic information system that is loyal to chemical thermodynamics. Microbes just know how to subsist through the least expense of biochemical energy. It is therefore good practice for the practitioner in bioremediation to weigh heavily any suggestion by vendors or operators that a site be treated via added organic materials, i.e., surfactants, emulsifiers, demulsifiers, etc. The possibility exists that these added compounds, aside from having the capability of lysing (killing) the microbial populations, might divert their metabolic attention away from the target contaminants by providing an alternate food source that is more desirable. If this should happen, the remediation will stall while the microbes are busy degrading the more attractive added compounds. It is best to detect this tendency during the bioassessment study where the treatment is an alternative under investigation (see Chapter 4), rather than to discover it during the full-scale field remediation.

There are three metabolic pathways by which microbes can reduce or alter hydrocarbon constituents: aerobic, anaerobic, and fermentative. The following is a brief discussion of each type of metabolism together with indications of how they can be applied in bioremediation technology.

Aerobic Respiration (O_2 as the final electron acceptor)

Aerobic and facultative anaerobic heterotrophs possess enzyme systems capable of oxidizing suitable substrates by transfer of electrons to molecular oxygen. Substrates that donate electrons are being oxidized, while those that accept the transferred electrons are being reduced. This is the most energy efficient of the three degradation pathways and is the mechanism of microbial choice where O_2 is available.

Substrate components in the waste matrix will be converted to CO_2, H_2O, and various other organic ligands (chemical species) utilized in the formation of new biomass and in biochemical assimilation and excretion within the respiring organisms. Generally, the reaction can be depicted as:

$$\text{Substrate} + O_2 \rightarrow \text{New Biomass} + CO_2 + H_2O$$

Many organisms that utilize aerobic respiration carry out incomplete oxidation of some substrates resulting in the formation of alcohols, aldehydes, and ketones. Many organic acids can also be produced, such as pyruvic, fumaric, citric, glycolic, oxalic, and a-ketoglutaric acids:

$$\text{Substrate} + O_2 \rightarrow \text{New Biomass} + CO_2 + H_2O + \text{Other Organics}$$

Whenever it is found that these organic reaction products tend to build up during biotreatment, this is an indication that the process is going only part way to mineral. Therefore, the process under investigation is performing a biotransformation rather than a complete biomineralization.

Historically, the field application of bioremediation technology has utilized the aerobic bacteria for the complete mineralization of organic contaminants. The reaction rates for aerobic metabolism processes are faster and the process is generally easier to control in the field when compared to anaerobic technology. Practitioners have simply analyzed the ongoing processes at a particular site and tried to enhance that activity. These days, it might be wise to investigate several metabolic alternatives before settling on a particular remedial design. Generally, if a site can be treated aerobically, without the necessity for changing and trying to maintain altered redox potentials, that might be the best course.

When halogenated organics are present, some aerobic processes may be designed that can cause destruction of many of these contaminants. Oxidation of some mono- and di-chlorinated organics proceeds via the substitution of a hydroxyl group and elimination of the hydrogen halide as in:

$$C_2H_4Cl_2 + HOH \rightarrow C_2H_4ClOH + HCl$$

The trihalogenated organics have been shown to undergo some sort of dehalogenation transformation in natural settings. However, completely halogenated compounds are not known (at present) to be amenable to aerobic dehalogenation. Compounds such as carbon tetrachloride and perchloroethylene are shown to be dechlorinated only through cometabolic reductive dechlorination in an anaerobic environment. Further discussion of bioremediation of chlorinated solvents will follow in this chapter.

Anaerobic Respiration (O_2 not the final electron acceptor)

Strict and facultative anaerobes are able to metabolize and grow in the absence of oxygen. Facultative anaerobes shift metabolic pathways to NO_3 as the final electron acceptor in the absence of molecular oxygen. Strict anaerobes become active only in the absence of O_2 by utilization of sulfate, carbonate, and specific organics as energy sources. These include the sulfate-reducing and methanogenic bacteria:

$$\text{Substrate} + NO_3 \rightarrow \text{Biomass} + N_2 + NO_2 + H_2O + CO_2$$

$$\text{Substrate} + NO_2 \rightarrow CO_2 + H_2O + N_2$$

$$H_2 + SO_4 \rightarrow H_2S + H_2O$$

$$\text{Organics} + H_2 \rightarrow \text{Biomass} + CH_4$$

$$2\,H_2 + CO_2 \rightarrow CH_4 + H_2O$$

Some microbes (*Pseudomonas denitrificans*) are able to convert NO_3 to N_2O and nitrogen gas. This conversion may be utilized for the *in situ* biological remediation of aquifers contaminated with nitrates.

Several groups of microbes are able to utilize sulfate as the final electron acceptor in the oxidation of organic substrates. The reduced end product of sulfate reduction is sulfide. Sulfite (SO_3) or thiosulfate (S_2O_3) can replace sulfate as an electron acceptor. Where oxidations are incomplete, organic acids are formed:

$$CH_3CHCOOH + SO_4 \rightarrow CH_3COOH + CO_2 + H_2S$$

Oxidation of molecular hydrogen occurs as:

$$H_2 + SO_4 \rightarrow H_2S + H_2O$$

In general, sulfate reducers require an environment of low redox potential (–100 meV and below) free of oxygen. Where isolated colonies occur in areas protected from free oxygen, even very slow metabolism produces small amounts of sulfide and growth conditions become more favorable. As the localized redox potential continues to fall, sulfate reduction becomes a self-reinforcing process and can result in rapid microbial growth. Initiation of growth can occur in any microhabitat from which oxygen is excluded, as occurs beneath a vigorous growth of aerobic organisms.

Where there is efficient removal of sulfide to prevent the inhibition of growth (escape of H_2S, utilization by aerobic sulfur oxidizers, or direct association with metals causing corrosion), the sulfate-reducing activity continues as long as the substrate is available. Anodic corrosion of ferrous metals occurs in this fashion.[5] In most cases, this anaerobic bacterial corrosion appears to be attributable to the genus *Desulfovibrio*. A motile, spore-forming sulfate-reducing genus is *Desulfotomaculum*. In addition to the common soil and aquifer microbial populations, some *halophilic* (salt-loving) species are known to be sulfate reducers.

In the context of bioremediation, there is an increasing use of anaerobic bacteria in field trials and their use in bioreactors has found wide acceptance (see Chapter 8). The characteristic problem faced by those attempting to apply this technology in the field is maintenance of a remediation zone free of O_2. In some cases, low-oxygen matrices have been found to exist in altered natural environments and aquifers.[6] Anaerobic field bioremediation was not a reality during the 1980s, but researchers (and indeed, even practitioners) may well demonstrate a level of competency during the 1990s. A case in point is the results obtained from field trials carried out at the Moffett Naval Air Station, California.[7] These trials gave evidence that *in situ* field anaerobic bioremediation of chlorinated solvents may indeed become a reality in the near future.

Fermentation

Yeasts, fungi, and certain bacteria have the ability to metabolize organic compounds through oxidation-reduction reactions to form less complex organics. In these metabolic reactions, both the electron donors and acceptors are organic compounds. Organic substrates are then only metabolized if they can be both oxidized and reduced. Several classes of the more complex organics (especially the carbohydrates) can be utilized during fermentation as they yield both oxidizable and reducible intermediates. Among the fermentable substrates are various sugars, polysaccharides, organic acids, biogenic excretions, and the lifeless cells of dead microorganisms. Products of fermentation are quite numerous and varied depending on the type of substrate, fermenting organism, amount of oxygen available, and environmental factors.

Fermentation reactions are especially varied among the many groups of fermenting bacteria. Among the end products of carbohydrate fermentation are a number of acidic and alcoholic compounds, carbon dioxide, methane, and hydrogen. Nitrogenous substrates will also yield ammonia, as is the case when the anaerobic sporeformers metabolize amino acids. Typical reactions yield carbon dioxide, ammonia, and 1-, 2-, 3-, and 4-carbon alcohols and acids.

In the context of this discussion, at the present time fermentation has not yet found wide use in environmental bioremediation. It is more likely to be encountered operating in a site under study for treatability assessment. In this instance, it is the fermentation products at the site that will have to be addressed during bioassessment. By establishing and maintaining aerobic conditions that may act as an inhibitor of this process, unwanted fermentations can sometimes be controlled or eliminated as an impediment to aerobic bioremediation of the site. However, there seem to be instances when active fermentation promotes the destruction of certain polycyclic aromatic hydrocarbons under actual aquifer conditions.[8] Recent advances in fermentation technology for *in situ* bioremediation are covered in Chapter 12. It is therefore anticipated that site remediation by selective fermentations should enhance the science and practice of bioremediation in the future.

BIOOXIDATION IN WASTEWATER TREATMENT

Perhaps the best known application of bioremediation is microbial aerobic biooxidation of wastewater. As discussed previously, biological treatment of wastewater for the destruction of organic contaminants is an ancient art that has been practiced for centuries. Basically, the process is accomplished through introduction of the wastewater to a properly acclimated microbial population in a carefully controlled environment containing appropriate nutrients. Domestic sewage and many industrial aqueous wastes are treated in this manner. The most efficient degradation designs incorporate aerobic microbes for the bioremediation of contaminated soils, sludges, and groundwater.

In terms of metabolic rates, aerobic bacteria are usually the more aggressive degraders compared to anaerobic and fermentative organisms. The preference for aerobes in the design and execution of bioremediation projects is commonplace, though not universal. The use of anaerobic bacteria has special application in the treatment of recalcitrant compounds and under special inhibiting situations. These are almost exclusively carried out in bioreactors, but field demonstration of *in situ* anaerobic bioremediation is a fact. These reactions will be addressed in the following chapters.

In common wastewater treatment plants, the biomass is suspended in the waste as free-floating floc and various methods are employed for control of these biological solids. Extended aeration, contact stabilization, activated sludge, oxidation ditches (race tracks), and aerobic digestors all employ suspended growth aerobic degradation for organics removal.

In some treatment applications the biomass is fixed on a plastic film, a permeable membrane, a semisolid or porous substrate, or some other foundation that requires that the wastewater flows over the biomass which remains stationary. An example of this arrangement is the trickling filter where the biomass adheres to smooth, medium-grade cobble. Packed towers, rotating biological contactors, and several other bioreactor configurations employ this method (see Chapter 8).

Treatment systems for contaminated soils and sludges must be designed and operated in a much different fashion than those which treat wastewater. Some parameters such as oxygen transfer, dissolved oxygen, nutrient levels, temperature, mass balance, etc. must be maintained as in wastewater treatment. However, when treating solids in this manner, the mass loading (percent solids) will be critical to horsepower requirements, aeration, mixing, and ultimate oxygen transfer. The temperature of the reactor solution may sometimes be of critical concern (such as when treating a native soil under less than optimum "native" conditions). Nutrient levels are much harder to maintain in a solids reactor and sampling protocols will need adjustment for observed conditions.

BIOATTACHMENT

In designing biological treatment schemes for contaminated soils, sludges, and groundwater, it is useful to understand that microbes which are active biooxidizers are normally attached to some substrate rather than free in the aqueous phase. They are commonly attached to the soil and sludge particles where they form colonies. If they become dislodged from a surface, they reattach in short order. In the treatment of aqueous waste where the microbiota is suspended in the waste, they simply attach to each other forming the familiar biological floc. The more soluble contaminants which remain in the aqueous phase are able to contact the microbes for immediate destruction. At the same time, the attachment phenomenon is fortunate for treatment of the less soluble organics in natural aquatic environments, especially aquifers. The less soluble organics are found adsorbed to soil particles and other solid surfaces. Being fixed on these surfaces, they are available for intimate contact when the colonizing microbes settle in and form a surface film over the solid particle. This condition is commonly found in aquifers that have been contaminated with the heavier, semivolatile or nonvolatile pollutants, including the hydrophobic solvents.

It is this surface attachment of degrader microbes that presents a delicate problem for bioreactor treatment of solids. In an attempt to increase mixing and aeration in the reactor by increasing the air flow, the resulting violent mixing increases shear forces that dislodge microbes attached to the solids. Even though DO measurements might indicate good treatment conditions, overall degradation efficiency suffers due to the dislodged biomass which is then no longer in contact with the sorbed insoluble organics. In this case, too much aeration and mixing can be counterproductive.

BIOSORPTION

Microbes possess the ability to provide protection for themselves against practically all forms of environmental threats to their health, and even their very existence. It is not conceivable that many unicellular life forms have gone extinct. This notion grows out of the discovery (over the many years of modern microbial research) of the existence of a myriad of protective mechanisms these fascinating life forms can evoke when threatened. Not only can many of them shift their metabolism to accommodate the changing physical and chemical environments and available nutrients and carbon sources, but they can produce protective enzymes and structures for long-term survival. One mechanism that can be exploited for use in bioremediation of contaminated environments is the phenomenon of *biosorption.*

Whenever a natural microbial environment is invaded by toxic metals, certain microbes are capable of secreting a "sticky" exocellular coating of protein, lipoprotein, or polysaccharide that is bound to the external surfaces of its cell membrane which attracts and binds the metal atoms. This action apparently serves to:

1. Provide a binding site that traps the metal atom from solution
2. Isolate the metal atom so that it is unavailable for transport across the membrane into the cell
3. Bring the metal atom into close contact with the cell for binding with other organic ligands that form organometallic complexes that can either detoxify the metal or provide a mechanism for mobilizing it away from the microbe

Whatever the case, microbes retain this extraordinary ability across widely separated taxa. The biochemical and genetic processes that initiate and terminate this highly specific activity are unknown. It is probable that this area of microbial response to environment will constitute a major arena for investigation in the future. At least one investigation at Los Alamos National Laboratory has addressed the phenomenon of bacterial accumulation and potential transport of transuranic metals.[9]

BIOSOLUBILIZATION

Colonizing microbes must transport the organic substrate molecules across the cell membrane into the cellular cytoplasm for digestion. If the organic is especially insoluble, then the microbes can secrete extracellular enzymes (for extracellular digestion of the contaminant) or surfactant *biopolymers* that can solubilize the pollutant for transport across the cell membrane for use as food inside the cell. By this method, adsorbed organics are degraded, albeit more slowly, than the more soluble contaminants. Production of microbial biopolymers and surfactants is regularly used in the petroleum industry for subsurface-enhanced oil recovery from depleted oil fields.[10]

It is precisely this microbial production of surfactants that accounts for one of the peculiar aspects of *in situ* bioremediation of groundwater. During acclimation of the native population of aquifer-dwelling microbes and soon after start-up of the remedial system, it is often the case that the operator will see increased levels of dissolved hydrocarbons in the monitoring samples taken at the extraction well (see Chapter 9). This will occur as a spike in the analysis for target compounds. Sometimes the effect is dramatic and can amount to an order of magnitude above observed baseline levels. The phenomenon is usually short-lived and returns toward observed pretreatment levels in a few days.

MICROBIAL DETOXIFICATION OF METALS

It has long been known that certain microbes can alter the redox potential or valence state of some of the metals. This information led to the development of processes for removal of heavy metals from wastewater.[11] Early understanding about the role of bacteria in enhancing the leaching of copper ores led to the discovery of microbial reduction and biosorption of metals. Today, the state of the art includes knowledge of the microbial ability to immobilize some metals and mobilize others, depending upon their physical behavior under aerobic and anaerobic conditions. In contrast to organic pollutants, metal contaminants are not biodegradable. However, many important toxic metal contaminants can be transformed to less-soluble or even volatile species by anaerobic bacteria.[12] Microbial volatilization of selenium has been reported by workers at the University of California, Riverside.[13] Soil levels of selenium were reduced 27% per year by microbial conversion to volatile dimethylselenide. Microbes employ several methods for reducing the toxic threat of metals:[4,14]

1. Adsorption to living or dead cell walls
2. Ion-exchange at the cell wall
3. Extracellular precipitation
4. Forming organometallic complexes

The metal U^{+6} is an important heavy metal groundwater contaminant at mining sites. This valence state of uranium can be bacterially reduced to the U^{+4} oxide mineral uraninite by either sulfate-reducing or dissimilatory iron-reducing bacteria:[15,16]

$$UO^{2(+2)} \rightarrow UO_2$$

Selenium and mercury can be reduced to their insoluble metallic forms and bacterial reduction of sulfate can reduce Zn, Pb, and Fe to their respective insoluble sulfides. Sulfate-reducing bacteria in aquatic and marine sediments reduce dissolved metals to insoluble metal sulfides. The iron sulfide in coal deposits is the result of bacterial sulfate reduction. This process can be utilized for the destruction of sulfates and immobilization of metals in a single step.[17-20] In contrast, insoluble $Pu^{(+4)}$ is bacterially reduced to soluble $Pu^{(+3)}$ through bioreduction.[15] There is also experimental evidence for potential natural bioremediation of heavy metal plumes in soils and groundwater.[21]

In these altered oxidation states mediated through microbial action, the metal chemistry is changed and often any metal toxicity is destroyed. A well-known example is the fact that, while soluble hexavalent chromium is extremely toxic, by microbial reduction to the trivalent form, chromium toxicity can be reduced or eliminated.[22] Microbes deal with environmental metal toxicants by reducing (or sometimes oxidizing) them to less-toxic or nontoxic forms; they just find the redox potential that renders the metal least toxic and adjust their metabolism accordingly. Cadmium and copper can be concentrated in biopolymers produced by exposed microbes.[23] It is instructive to examine the mechanics of metal metabolism starting with the well-known iron-metabolizing lithotrophic microbes.

Lithotrophic aerobes have the ability to derive energy from the oxidation of inorganic compounds such as hydrogen gas, iron, and reduced compounds of nitrogen and sulfur. They must obtain their carbon for growth from CO_2 and can develop in environments almost entirely devoid of organic matter:

$$CO_2 + H_2 + NH_3 + H_2S + Fe^{+2} + O_2 \rightarrow \text{Biomass} + H_2O + NO_{2,3} + SO_4 + Fe^{+3}$$

The use of these microbes is effective in reducing soluble concentrations of nitrates and sulfates in aquifers. As an example of microbial alteration of metal redox potential, normal iron metabolism can lead to bacterial corrosion of ferrous metals and ores. Virtually all organisms require iron as an essential trace element in metabolic enzyme formation. Some of the lithotrophic bacteria, however, are capable of deriving all their energy requirements from the oxidation of ferrous iron (soluble Fe^{+2}) to ferric iron (insoluble Fe^{+3}):

$$4Fe^{+2} + 4H^+ + O_2 \rightarrow 4Fe^{+3} + 2H_2O$$

In so doing, these microbes also require CO_2 (either gaseous or dissolved as CO_3^{-2}, HCO_3^-, or H_2CO_3) as a carbon source for biosynthesis. This metabolic pathway operates in members of some 20 bacterial genera, including *Ferrobacillus, Gallionella, Leptothrix, Thiobacillus,* and at least four fungi.

Iron can also be solubilized from metallic or ferric iron to the soluble ferrous form by several mechanisms. Corrosion of iron and its alloys can occur as pitting corrosion under a scale deposit or under a colony of sulfate-reducing anaerobic bacteria. Thus, the hydrogen sulfide formed reacts with iron to form the disulfide:

$$SO_4 + 5\ H_2 \rightarrow H_2S + 4H_2O$$

$$Fe + 2H_2S \rightarrow FeS_2 + 2H_2$$

The FeS_2 produced by this reaction can be solubilized by sulfur-oxidizing bacteria, liberating the soluble ferrous iron:

$$FeS_2 + H_2O \rightarrow Fe^{+2} + H_2SO_4$$

Corrosion can also occur as an indirect result of bacterial metabolism. Where a reducing environment is generated when aerobic microbial growth depletes dissolved oxygen, metallic iron is dissolved at anodic corrosion sites on the metal or ore surface. The corrosion cell becomes active beneath actively growing bacterial colonies and results in tubercle formation and pitting corrosion.[5] A similar mechanism has been discovered by which microbial corrosion of aluminum aircraft fuel tanks takes place beneath colonies of sulfate-reducing and associated bacteria. Accumulated condensation of water in the tank bottom provides the microbial habitat. Also problematic for municipal and industrial systems containing water are sulfur-oxidizing microbes that actively corrode iron pipes and can generate dangerous quantities of H_2SO_4.

Some workers have identified a mechanism by which ferric iron is biologically reduced to ferrous iron by direct enzymatic action. Organisms of the genus *Bacillus* and others are active in the reduction of iron, although the exact mechanisms remain to be discovered. Some organisms are also capable of transforming the redox potential of other metals, such as manganese and uranium. Acidic solutions containing Cu, Cd, and Zn can undergo biotreatment in which these toxic ions are captured through biosorption onto immobilized yeast cells.[24] The active detoxification of uranium via bioremediation of metal-contaminated groundwater from a DOE site in New Mexico was demonstrated in the laboratory.[25] Facultative anaerobes can be shown to alter redox potential of toxic metals to nontoxic states that may be more soluble in moving groundwater (down-gradient dilution playing a major role in dissipation of the threat). If the solubility of a still-toxic metal is reduced, the consequent immobilization of the pollutant might be an acceptable end result. This concept is ready for an initial field trial in an actual aquifer remediation.[26]

HALOBACTERIA

The ability of brackish and saltwater microbes to actively degrade a wide variety of hydrocarbons has been well established and is easily demonstrated.[27] This fact is especially useful in remediation of maritime catastrophes and sites where petroleum and petrochemicals contaminate salty soils and water. Those sites where the groundwater is saline, or soils are high in salt content, are amenable to bioremediation of organic contamination. Some sites that are adversely affected by high salinity may benefit from some means of brine or salinity reduction, if this is, in fact, a major problem. If the native microbes do not seem to be sufficiently acclimated, dilution by freshwater or soil washing can reduce the high salt content in some sites.

The ability of bacteria to adapt to life in highly saline environments is not limited to any special microbial class. Representatives from several branches including many soil microbes have halophilic types or can acclimate to hypersaline conditions. Aside from the classic halophilic family of microbes, the *Archaebactereacae* (*Halobacteracae*, *Halococcus*, etc.), there are many common soil bacteria (*Pseudomonas, Bacillus, Micrococcus, Arthrobacter, Vibrio,* etc.) that are readily adaptable to hypersaline environments. Several common bacteria utilized in activated sludge treatment of soluble organic chemicals have been shown to be active at salinities up to three times that of seawater (10%). Some bioremediations must be performed in highly saline sites. Speedy adaptation of nonhalophilic microbes, when introduced into a highly saline environment, has been well established.[28,29]

Two common soil microbes show particularly amazing adaptability to highly saline environments. The aerobic bacteria of genus *Vibrio* and *Pseudomonas* comprise 60 to 70% of microbes isolated from coastal salterns with salinities greater than 25% TDS (total dissolved solids). As salt concentrations approach saturation, microbes of genus *Alteromonas, Alcaligenes,* and the halophilic archaebacteria *Halobacterium* that are present will begin to metabolize.

Of the nine major groups of halophilic bacteria, most are aerobic, require salt molarities of 2.5 to 5.2 (which translates to 15 to 35%), exhibit maximum growth at 45 to 50°C, and have a variety of nutritional requirements. Growth of *Halobacterium sodomense* is enhanced by the presence of clay minerals. Extreme halophiles possess the ability to grow in oxygen-poor (microaerophilic) environments and are involved in putative fermentation in the absence of oxygen. Some members seem to be oligotrophic (able to metabolize at extremely low levels of available organic carbon).

Although the extreme halophiles are generally aerobic, some possess alternative modes of respiration. Several species can grow anaerobically, some are methanogenic, some are sulfate reducers, and some are able to ferment organic substrates. Some of these anaerobes have demonstrated a potential for glucose fermentation and cellulose decomposition. Thus, there is a diversity of halophilic anaerobes, and in addition to the many strict aerobes, there are many microaerophilic halophiles able to grow on extremely low oxygen concentrations.[30]

MIXED POPULATIONS

Natural and polluted environments are quite different from those present in the laboratory. In contrast to the pure culture and single substrate systems of classical microbial investigation, the field environment (an open system where the art and science of site remediation is performed) is extremely varied and complex. In virtually no circumstance is a single microbial species responsible for the degradation and destruction of pollutants in the field.

Most natural environments are simply loaded with microbes. Even a heavily polluted site contains multiple microbial species and many of these may have become acclimated to existing contaminants. It is common for these mixed populations to act in concert in metabolizing and degrading the organics available through the concept of *commensalism*. Sometimes the species available cannot completely destroy a contaminant of interest. The target organic is degraded to a certain point and left there. Whenever a site characterization shows that there is an accumulation of these intermediate metabolites, it might indicate the lack of a class of microbes that would be beneficial to that site remediation. The question of whether to add the "missing link" microbes in the degradation pathway lies in whether or not the intermediates are harmful. If the end point of the remediation can tolerate the presence of the intermediate, then the remediation might be accomplished by biostimulation of the indigenous species present at the site. If the intermediates are not tolerable, then inoculation (bioaugmentation) may be considered or some other remedial approach may be warranted.

COMETABOLISM

This term, introduced into the vernacular of environmental remediation in 1981,[30] has a specific meaning that is very often misapplied and commonly misunderstood. Degradation by a *mixed population* in which two or more organisms acting together effect the stepwise destruction of a specific compound is *NOT* cometabolism. The authors have seen this term improperly used to describe degradation by a mixed population, a concept in which organism A causes the partial destruction of a compound, but cannot degrade it further. The end-product metabolite of organism A is then attacked by organism B (and possibly C, D, and E in a series), which carries out the remainder of the degradation to yield a nontoxic end product through a process termed *commensalism*. Therefore, degradation by two or more organisms or a mixed population does not constitute cometabolism.

The generally accepted definition of cometabolism is limited to *microbial action that modifies chemical structure without yielding energy utilized for growth* of the organism involved. In other words, the cometabolized compound is not utilized as food for the organism. In fact, the concept is actually applied to the substrate rather than the organism. As they are not useful to the organisms involved, the microorganisms and metabolic pathways that are commonly employed in cometabolic degradation of environmental pollutants are not even necessarily *biochemically aware* of the target compounds that the site manager is trying to remediate. For illustrative purposes, let's assume we have a site contaminant we have targeted for destruction that cannot easily be degraded through microbial action (sad, but sometimes true). It is often the case that recalcitrant compounds (those not degraded for energy and growth by microbes) are considered by many to be nonbiodegradable...*directly.* However, through experience, we might know that the target compound can be destroyed *indirectly* by cometabolic action of certain organisms by simply providing a *primary substrate*. In other words, by enticing the microorganisms to attack an *alternate food source* (the primary substrate that we will add to the site), the metabolic machinery that accounts for the degradation of the primary substrate will accidentally (but not for us, because we designed it this way) and fortuitously also destroy the target compound we wish to degrade.

In order to proceed with our plan, we would want to bioassay the contaminated site media for those organisms capable of carrying out the cometabolic remediation. Finding they are present (if not, we might bioaugment or inoculate the site), we now have a possible route for successful bioremediation of the site. If these needed organisms can be stimulated through addition of required nutrients (a proper primary substrate) and control of the environment, then we might see that the target compound is reduced below action level concentrations. Of course, this must be confirmed in a proper laboratory bioassessment investigation before applying the technology in the field.

In the real world of environmental site cleanup via bioremediation, aerobic cometabolism was discovered to operate in aquifers contaminated by chlorinated solvents.[3] Trichloroethylene (TCE; widely thought to be nonbiodegradable) was found to degrade through *in situ* microbiological cometabolism. Wherever a group of organisms called methanotrophs (methane degraders) is found in an aquifer, cometabolism of chlorinated solvents is a possibility. Methanotrophs oxidize methane for carbon and energy through the action of a metabolic enzyme called monooxygenase. This enzyme is the destructive biochemical catalyst for methane oxidation. As it happens, the same enzyme also catalyzes the destruction of TCE, although this degradation provides no benefit to the methanotroph. TCE is degraded in this fashion, but it is actually being *cometabolized* along with the methane and provides the organism with no benefit. The organism is not metabolizing the TCE, but through biological action the TCE is degraded. A similar reaction is responsible for the cometabolic destruction of toluene in a bioreactor.[31] The same phenomenon operates during the microbial reduction of chromate.[22]

CHLORINATED HYDROCARBONS

Halogen-substituted organic compounds constitute one of the greatest threats to the subsurface environment and our groundwater resources. Much scientific effort has been expended toward finding economical means of destruction or removal of these toxic and sometimes carcinogenic compounds.

For many years chlorinated solvents were thought to be nonbiodegradable. Then researchers began finding their degradation products in soils and aquifers. This led workers to begin a systematic search for the organisms and specific degradation pathways that were at work. Now it is known that several of the more common halogenated solvents undergo microbial biodegradation in the natural environment. It is presently thought that the fully chlorinated solvents are biologically degraded from the perchloro to the less chlorinated compounds only through cometabolic reductive dehalogenation by anaerobes (which also requires the presence of a cometabolic primary substrate). This removal of the first halogen atom seems only to occur in the presence of anaerobes at low redox potential. Subsequent degradation can occur in the anaerobic pathway, but may *stall* at some point before complete mineralization. Another anaerobic pathway that seems to be effective in destroying some chlorinated compounds is *denitrification*.

On the other hand, *aerobic* degradation through cometabolism has been shown to degrade these less-halogenated intermediates all the way to mineral (CO_2, water, and free chloride). Methane-oxidizing bacteria can destroy chlorinated solvents through generation of a cometabolic enzyme. However, it is sometimes difficult to supply sufficient oxygen into the subsurface to support this type of remediation. Nitrates, sulfates, and carbon dioxide can be utilized by certain microbes as alternate electron acceptors where oxygen is unavailable in high concentrations, or when its use is impractical. A thorough discussion of the current technology for bioremediation of halogenated hydrocarbon solvents in soils and groundwater was published in 1994.[3]

RADIATION EFFECTS

In nature, microorganisms are exposed to continual low dosages of ionizing radiation originating from terrestrial (radionuclides in rock, soil, air, and water) and extraterrestrial (cosmic ray) sources. One of the consequences of radiation — genetic mutation (which results in the loss of genetic information carried in the DNA) — can limit the environmental endurance and choice of food substrates available to the mutated microbes. Mutation can be shown to have only two effects on organisms: either it does nothing observable (alters a nonessential portion of a DNA sequence), or it causes a defect in an essential DNA portion (gene) and results in a defective enzyme or other protein which can alter the ability of the organism to survive in its present environment, or prevents metabolism of the substrates it depended upon for food previous to the mutation. So, in short, mutations are harmful because of their alteration of the information carried in the DNA. It is sometimes stated that *beneficial mutations* are the driving force for natural selection, implying that biological organisms are continually "improving." The truth is, we have never observed any of these hoped-for beneficial mutations, but rather we have documented tens of thousands of harmful ones. Neither have we ever witnessed the creation of a new species. If allowed to accumulate, mutations build up to a damage potential (the genetic load) that can result in the extinction of a species (another well-documented fact of natural history).

The destructive effects of mutation can often be overcome through a mechanism called *DNA repair.* Through an organism's ability to perceive that damage has occurred to its DNA, the process of repair then assesses the damage, cuts out the mutated sequence of DNA, synthesizes a correct sequence, and inserts it in its proper place to complete the repair. In this fashion, organisms can restore lost or altered information caused by mutation.

The natural selective pressure of environment is what moderates the destructive effects of radiation through the development of protective mechanisms such as acclimation in microbes for defense against radiation damage. Therefore, in any contemplation of bioremediation of sites contaminated by radiation (including *radioactive mixed waste*), the only real consideration will be the types and varieties of organic materials and toxic metals involved, because the radiation type and level in typical mixed wastes will be of no real consequence to the microbes.[9]

From a practical point of view, generally, as radiation exposure increases, so does the frequency of lethal mutations which selects against the nonresistant organisms. As less resistant organisms die off, the more resistant strains multiply to fill the open niche and gain competitive advantage. As a result of this phenomenon, the use of radiation for food sterilization raises questions concerning the induction or selection of radioresistant microbes; a possibility that could be counterproductive.

ACQUIRED RADIATION RESISTANCE

One of the most radiation-resistant bacteria ever isolated (*Micrococcus radiodurans*) was discovered in canned meat which spoiled in spite of having been exposed to several megarads of gamma radiation. This organism can withstand single doses as high as 500 krad without showing any significant inactivation and has been the subject of considerable research.

Radiation exposures of high intensity are much more effective as selective agents than those of low intensity. Under the selective pressure of radiation, resistant strains have been isolated from species of the genus *Escherichia*, *Salmonella*, *Achromobacter*, and a yeast. Even in organisms that are normally sensitive to radiation, the trait of radiation resistance can be induced or magnified by long-term exposure and the resulting acclimation of the organism.

In the course of experimentation with exposure to radiation, several additional radiation-resistant groups of microbes have been identified. Members of the nuclear industry were surprised in 1960 to find a nuclear reactor primary cooling water system which had extensive fouling caused by slime formations of a species of *Pseudomonas* bacterium. A ubiquitous, very adaptive, and prolific group of organisms, *Pseudomonas* species are normally radiation sensitive. However, this bacterium had become acclimated to the high radiation environment of the reactor where the average dosage was probably in excess of several hundred *rads*. When the damaged Three Mile Island #2 Unit reactor core was being defueled, at least ten species of bacteria were isolated from the core cooling water. These microbes had grown accustomed to a radiation-absorbed dosage level of from 1000 to 10,000 rad/hr.[32]

Of particular interest in the search for ever more resistant strains of bacteria through artificial selection is *Micrococcus radiophilus*. Isolated in 1973 from a sample of radiation-sterilized fowl, this organism shows ever-increasing radiation tolerance with certain strains surviving after exposure to single doses exceeding 1.5 mrad. In comparison, a lethal human dose of radiation is considered to be equal to about 600 rad.

The effects of radiation on bacterial adaptation were shown in an analysis of the soil microbes in close association with Los Alamos National Laboratory simulated organic waste. Bacteria isolated from radioactive waste burial site soil samples containing beta and gamma activity had more radioresistant isolates in the population than microbes from soil lacking detectable radioactivity. Studies showed that the bacteria isolated from transuranic (TRU) waste burial sites exhibited levels of radioresistance intermediate to that of *Bacillus subtilis* and *Micrococcus radiodurans*.

RADIOACTIVE MIXED WASTE TREATMENT

The foregoing discussion is illustrative of the high potential of bioremediation for the treatment of *radioactive mixed waste*. Bioprocessing treatment options have been explored for:

1. The destruction of the RCRA hazardous fraction
2. The immobilization or concentration for removal of the offending radionuclide

In sites where mixed waste is encountered, the remediation specialist might be able to employ the techniques described in this book to remove the hazardous fraction of the mixed waste, thereby transforming it to one of the designated NRC or DOE classes of RAD waste for disposal. Removal of the RCRA fraction allows disposal as a RAD waste under AEA regulations. In some cases, it may be possible to biologically treat the metallic portion of mixed waste through altered oxidation states, i.e., the detoxification of a metal, or perhaps altering its solubility in surface or groundwater or by specific accumulation and separation. Although this approach might not necessarily destroy the mixed waste, it may result in a more environmentally acceptable form of the waste.

Bioremediation of the various forms of mixed waste can sometimes be applied to degrade or transform:

1. The PCB or pesticide fraction of TSCA mixed waste
2. The RCRA fraction of mixed waste contaminated soils, sludges, and groundwater
3. The hydrocarbon fractions of scintillation cocktails, contaminated lubricants, turbine oils, hydraulic fluids, uranium processing and enrichment wastes, complex nitrates, and explosives residues

As mentioned earlier, biological treatment applied to these categories of mixed waste may allow complete destruction of the RCRA organic fraction, rendering the waste a simple NRC-regulated RAD waste for disposal. Another outcome could be the separation of the wastes for individual disposal, or at least a significant volume reduction to simplify storage for later disposal. When applied *in situ,* bacteria can often render suitable treatment for soils, sludges, and groundwater. Planting a cover crop for concentration of soil nuclides constitutes the practice of *phytoremediation.*

Mechanisms known to operate under conditions favorable to bioremediation where mixed waste is present include:

1. Bioadsorption of metals and nuclides to the exterior cell membranes of bacteria and algae, sometimes forming insoluble complexes with carbonates, nitrates, sulfates, phosphates, or silicates
2. Bioconcentration of these metals within the bacterial cells, or in cover crop vegetation
3. Alterations in redox potential or valence states of metals causing changes in toxicity, reducing solubility or mobility
4. Mobilization of toxic metals through complexing or chelation with enzymes or surfactants that are excreted by bacteria into their fluid medium

These treatment strategies can be applied to surface soils in place, *in situ* for soils and groundwater, or in bioreactor situations where complete control of the treatment parameters is possible. These reactions may facilitate reductions of risk by reducing the probable exposures to receptor organisms, or may reduce the toxicity of a waste. However, caution must be exercised when considering the biomass produced. It may become a health hazard due to toxic chemical contamination or radiation.[33]

Bioprocessing of mixed waste has been achieved in a pilot-scale sequencing bioreactor application for scintillation fluids. A strain of the aerobic bacteria *Pseudomonas* was found to destroy the p-xylene hydrocarbon fraction of these wastes in a patented process that treats both the liquid and the vapor phase. The same organism was found to tolerate or degrade an entire host of hazardous materials. When employed for plutonium-contaminated waste treatment, these microbes could incorporate 96% of the radionuclide into the cellular biomass. Similar processes have been developed for chlorinated aliphatic hydrocarbons (CAHs) in mixed waste, biofilters for vapor phase treatment, destruction of cyanide wastes, and for the biodecontamination of surfaces.[32]

CONSTRUCTED WETLANDS

Constructed wetlands are another important application in wastewater biooxidation by fixed-growth microbes. These dynamic and versatile surface treatment systems are effective for domestic and municipal wastewaters, urban runoff, agricultural wastes, industrial effluents, landfill leachate, coal pile and ash pond seepage, and acid mine drainage. They are simple systems employing soils, aquatic plants, and attached microbes for adsorption, assimilation, and destruction of organic and inorganic wastes. Because these systems have been covered elsewhere, the reader is referred to that source for further discussion.[34]

REFERENCES

1. Rochkind, M.L., Blackburn, J.W., and Saylor, G.S., 1986, Microbial Decomposition of Chlorinated Aromatic Compounds, USEPA, Report EPA/600/2-86/090, Cincinnati, OH.
2. Gibson, D.T., ed., 1984, *Microbial Degradation of Organic Compounds,* Marcel Dekker, New York.
3. Norris, R.D. et al., 1994, *Handbook of Bioremediation*, Lewis Publishers/CRC Press, Boca Raton, FL.
4. Means, J.L. and Hinchee, R.E., 1994, *Emerging Technology for Bioremediation of Metals*, Lewis Publishers, Boca Raton, FL.
5. *BETZ Handbook of Industrial Water Conditioning*, 1985, BETZ Laboratories, Inc., Trevose, PA.
6. Nuttall, H.E., Lutze, W., and Barton, L., 1996, Preliminary Screening Results for *In Situ* Bioremediation, TVI/HSRC Conference on Bioremediation of Mixed Waste, Albuquerque Technical Vocational Institute, Albuquerque, NM.

7. Semprini, L., 1996, Bioremediation of Chlorinated Solvents, Third TVI/HSRC Conference on Bioremediation, Albuquerque Technical Vocational Institute, Albuquerque, NM.
8. Grbic-Galic, D., 1990, Methanogenic transformation of aromatic hydrocarbons and phenols in groundwater aquifer, *J. Geomicrobiol.,* 8, 167–200.
9. Kliewer, G., 1996, Waste isolation pilot plant bacteria studied for contamination scenarios, *Los Alamos Newsbulletin,* 16(31).
10. Prescott, L.M., Harley, J.P., and Klein, D.A., 1993, *Microbiology,* 2nd ed., Wm. C. Brown, Dubuque, IA, p. 905.
11. Brierley, C.L., Brierley, J.A., and Davidson, M.S., 1989, Applied microbial processes for metals recovery and removal from wastewater, in *Metal Ions and Bacteria,* Chapter 12, Beveridge, T.J. and Doyle, R.J., Eds., John Wiley & Sons, New York.
12. Lovely, D.R., 1995, Bioremediation of organic and metal contaminants with dissimilatory metal reduction, *J. Ind. Microb.,* 14.
13. Frankenberger, W.T. and Karlson, U., 1988, Dissipation of soil selenium by microbial volatilization at Kesterson Reservoir, in *Hazardous Waste Treatment by Genetically Engineered or Adapted Organisms,* HMCRI, Washington, D.C.
14. Golab, Z., Orlowska, B., and Smith, R.W., 1996, Reduction of lead and uranium by *streptomyces* sp., *Water, Air, Solit Poluton,* 60.
15. Fredrickson, J.K. and Gorby, Y.A., 1996, Environmental processes mediated by iron-reducing bacteria, *Current Opinion Biotechnol.,* 7(3).
16. CeRaM, 1996, New Tools for Bioremediation: Groundwater Treatment at UMTRA Sites in Tuba City, AZ and Shiprock, NM. Status Report on Technology Development, Center for Radioactive Waste Management, University of New Mexico, October.
17. Barnes, L.J. et al., 1991a, Microbial removal of heavy metals and sulfate from contaminated groundwaters, in *Emerging Technology for Bioremediation of Metals,* Means, J.L. and Hinchee, R.E., Eds., Lewis Publishers, Boca Raton, FL.
18. Barnes, L.J. et al., 1991b, A new process for microbial removal of sulfate and heavy metals from contaminated waters extracted by a geohydrological control system, *Chem. Eng. Res.,* 60.
19. Barnes, L.J. et al., 1992, Simultaneous microbial removal of sulfate and heavy metals from wastewater, *Trans. Inst. Min. Metall.,* 101, C183–189.
20. Scheeren, P.J.H. et al., 1992, New biological treatment plant for heavy metal contaminated groundwater, *Trans. Inst. Min. Metall.,* 101, C190–199.
21. DeRosa, G., 1997, Experimental Evidence for Potential Natural Remediation of Heavy Metal Plumes, WERC/HSRC '97 Joint Conference on the Environment, Albuquerque, NM.
22. Bader, J.L., Gonzales, G., Goodell, P.C., Pillaiand, S.D., and Ali, A.S., 1996, Bioreduction of Hexavalent Chromium in Batch Cultures Using Indigenous Soil Microorganisms, HSRC/WERC Joint Conference on the Environment, Albuquerque, NM.
23. Fukushi, K. and Ghosh, S., 1996, Removal and Recovery of Copper and Cadmium by Aerobic Microbial Cultures Grown with Biopolymer Stimulators, HSRC/WERC Joint Conference on the Environment, Albuquerque, NM.
24. Wilkins, E., 1996, Bioremediation in a Gel Barrier Using Immobilized Yeast for Heavy Metal Removal, HSRC/WERC Joint Conference on the Environment, Albuquerque, NM.
25. Thombre, M., Thomson, B.M., and Barton, L.L., 1996, Microbial Reduction of Uranium Using Cellulosic Substrates, HSRC/WERC Joint Conference on the Environment, Albuquerque, NM.
26. Nuttall, H.E., 1997, University of New Mexico, personal communication.

27. Atlas, R.M. and Bartha, R., 1972, Degradation and mineralization of petroleum in seawater: limitation by nitrogen and phosphorus, *Biotechnol. Bioeng.,* 14, 309–317.
28. Rodriguez-Valera, F., 1988, *The Halophiles*, CRC Press, Boca Raton, FL.
29. King, R.B., 1989, Preliminary Microbiological Survey of the Underground at the Waste Isolation Pilot Plant, Chapter 3 of the Brine Sampling Evaluation Project annual report for Westinghouse Electric Corp., Carlsbad, NM.
30. Alexander, M., 1981, Biodegradation of chemicals of environmental concern, *Science,* 211, 132.
31. Liu, T., Fukushi, K., and Ghosh, S., 1996, Anaerobic Biodegradation of Toluene in a Plug-Flow Digester, HSRC/WERC Joint Conference on the Environment, Albuquerque, NM.
32. Wolfram, J.H., 1996, Bioprocessing Scenarios for Mixed Waste, TVI/HSRC Conference on Bioremediation of Mixed Waste, Albuquerque Technical Vocational Institute, NM.
33. Graves, D., 1996, Mixed Waste Bioremediation, TVI/HSRC Conference on Bioremediation of Mixed Waste, Albuquerque Technical Vocational Institute, NM.
34. Hammer, D.A., 1989, *Constructed Wetlands for Wastewater Treatment*, Lewis Publishers, Chelsea, MI.

CHAPTER 4

The Bioremediation Laboratory

The art and science of bioremediation can present quite different views of the challenges involved, depending on who is looking. From the point of view of the novice, things may (1) look awesomely complicated, and the probability of success on a given project may seem slim, or (2) look very simple, with success virtually assured with "No worries, Mate." However, those views are limited by inexperience. In the first instance, it is likely that too much effort and/or money might be expended for no good reason if the site data indicate no justification for the expense. Some projects are completed in a timely manner with only a minimum of testing. In the second case (and this is where most newcomers go terribly wrong), many times a project is initiated with little or no (or improper) data to guide a proper remedial design, and failure is assured.

The authors talk with potential clients on a regular basis who, as site owners, have watched the calendar pages turn *ad infinitum* with little evidence of progress toward closure. Many times these projects have either stalled or never gotten off the ground due to the lack of some very rudimentary site or monitoring data. At a specific site in Arizona, site soils had been undergoing "so-called" bioremediation for almost 3 years and had stalled at 20 times the regulatory limit for heavy fuel oil. Upon review of the site monitoring data, it was discovered that the sampling and analysis protocol had neglected to test for the correct microbial nutrients (testing had been ordered for total Kjeldahl nitrogen instead of ammonia nitrogen). As a result, fresh samples were analyzed and showed that ammonia nitrogen was at zero; the remediation could not progress without nutrient additions and the work plan had specifically stated that "nutrient amendments would not be needed." This remediation had been stalled for 30 months, and the contractor (a billion dollar revenue, 50-year-old multinational environmental engineering firm) did not know why! When the correct nutrients were provided in a proper amendment, the soils cleaned up in short order. In another similar case in New Mexico involving petroleum contamination, the project had stalled because site soils had simply gone dry in the arid southwestern climate; moisture additions through regular sprinkling revived the natural (already acclimated) microbial population to complete the job.

In reality, those who have attained some degree of expertise in this field may find that some projects can be approached with confidence and with very little preparatory investigation. For instance, faced with the job of remediating freshly excavated soils containing volatile fuel hydrocarbons, an experienced bioremediation consultant might design and execute a simple soil pile or land treatment scheme. Still, she will make the necessary analyses to ensure that the soils contain adequate population numbers of acclimated microbes able to degrade the site contaminants in a timely fashion, and that the required nutrients and moisture are applied in the most cost effective and advantageous manner. This all might seem unnecessary to the novice, but it could mean the difference between success and failure of the project. If the regulatory authorities are monitoring the progress of the project, it behooves all concerned to get an early closure at the least expense, especially when tax monies are being expended. In other words, there is a minimum of testing and investigation that should take place with any project; only an experienced consultant will know when to skimp on the budget.

So then, there are times when a full treatability study should be pursued, and times when it might be abbreviated. When site conditions suggest the possibility of difficulties, a greater depth of inquiry is warranted.

Not all biological treatment system designs and contemplated bioremediation activities require extensive up-front laboratory testing as part of the phased approach to successful bioremediation. However, *when warranted*, these tests serve a number of functions associated with the project, including:

1. The provision of valuable design information
2. A determination of existing conditions at a site
3. A search for potential "knock out" factors which might preclude the use of bioremediation
4. An initial definition of a proposed biological treatment process
5. A basis from which to determine the probability of success

Laboratory testing is sometimes erroneously looked upon as an unnecessary delay in a project, a "re-inventing of the wheel," or the display of a lack of confidence or competence on the part of the professionals who are recommending a biological treatment approach. These are unfortunate misconceptions based on ignorance and indifference to acceptable bioremediation practice. The design and conduct of a quality laboratory study (or even a minimum effort) can save valuable time on a project and can significantly *reduce the ultimate cost.*

Even though bioremediation practice involves the talents of many technical professionals in the conduct of a proper field remediation project, perhaps the most important individual at this point in the project is the biotreatment microbiologist. The laboratory setup, equipment, and procedures for the following testing protocols must be in accordance with accepted standard environmental microbiology. In addition to the steps outlined in this text, copies of Standard Methods (APHA), EPA Manual SW-846, and a good microbiology lab manual should be on hand for ready reference. Plate counts will be needed for total heterotrophs and specific contaminant degraders. The *broth* and *agar* media required can be blended in the lab or purchased from reputable suppliers. Special apparatuses might sometimes be required, but in many cases the modestly outfitted facility will suffice.

It is useful to divide laboratory studies into two categories in discussing protocols and objectives:

1. Bioassessment screening studies
2. Biotreatability studies

These terms are often wrongly used interchangeably. Other common terms in use are feasibility studies, optimization studies, preliminary engineering studies, and site assessments. The following discussion is given in order to present the key concepts supporting the initial laboratory testing and how it relates to the preliminary information needed for system design and monitoring protocols. *Never assume* that an environmental laboratory can perform the testing protocols outlined herein, even though their Statement of Qualifications (SOQ) says they can. You must go and **see for yourself**; talk with the technicians and especially their microbiologist (*chemists* are often tagged for this work by the lab manager, and in that case — beware!).

BIOASSESSMENT SCREENING STUDIES

Bioassessment is a set of screening tests that are designed to *assess* whether *biological treatment* can be utilized at a particular site. Screening tests compare initial site data to a known set of judgment criteria. A screening study for bioremediation compares the results of a series of physicochemical and microbiological indicator tests (performed on samples obtained from the site under study) to those factors known to be required for successful bioremediation of similar contaminants and site-specific parameters. These tests are relatively simple to conduct and can be readily performed in 5 to 8 weeks. Under most situations, these results lead to a *"go" or "no go" decision* for bioremediation as a viable treatment option for a specific site.

Screening studies are designed to determine whether biodegradation of the target contaminants occurs when a suitable population of microbes contacts contaminants under an appropriate set of site conditions. These conditions include the proper balance of nutrients, sufficient moisture, the required electron acceptor in sufficient supply (oxygen in most cases), and a conducive environment. These tests seek to identify problem factors that would inhibit or prevent the use of biological treatment. They are also geared to predict the efficiency of nutrient transport through the contaminated matrix or the engineered treatment system.

The screening study is typically implemented after initial data from a site characterization study are reviewed and some idea of the extent of contamination is known. Site characterization data will indicate the nature of contaminants, their location and areal extent, their concentrations and mobility, their probable fate if untreated, and the chemical, physical, and biological conditions at the site. With this information in hand, the proper sampling and testing protocols can be prepared to ensure maximum usefulness of the data. Screening studies must consider all impacted soil and groundwater from a site, as well as any required makeup water, process water, or sludge that may be present or that may be produced during treatment. It is an exception, rather than the rule, that only a single matrix will be examined in a screening study. Whenever groundwater is involved, both soils and groundwater samples must be analyzed. Typical volumes of solid matrix samples gathered for a study range between 1 and 20 kg. Typical sample volumes for liquid matrix collected range from 1 to 10 l.

Costs for a bioassessment screening study for an individual site can range from $1000 to $10,000 depending on the number and volume of samples to be tested, the analytical turnaround time required, the actual number of tests conducted, and the complexity of reporting requirements. Most laboratories specializing in these studies offer them at a fixed fee, although some still charge on a time-and-materials basis.

The most common physical and chemical parameters evaluated in a screening study for bioremediation include:

WATER

pH
Hardness (Ca, Mg)
Metals (Fe, Cu, Mn)
Dissolved oxygen
Dissolved nutrient content (P, N)
Specific organic carbon contaminants (TPH, O&G, BTEX, etc.)
Total organic carbon
Biochemical oxygen demand
Chemical oxygen demand
Total suspended solids

SOIL and SLUDGE

pH
Metals (Fe, Cu, Mn)
Total Organic Carbon
Nutrient content (P, N)
Specific organic carbon contaminants (TPH, O&G, BTEX, etc.)

MICROBIOLOGICAL ASSAY

In addition to the specific testing described above, it is necessary to ascertain the microbial health of all media (soil, sludge, and water) to be remediated. This requires analysis to determine the relative numbers of viable microbes and the extent to which they can utilize the target compounds for growth and metabolism.

The following tests should be considered a minimum for bioassessment screening:

1. Microbial nutrient content:
   Total phosphorus and ortho-phosphate
   Total nitrogen (Kjeldahl) and ammonia nitrogen
2. Nutrient interactions (this should identify any precipitation of ortho-phosphate and/or ammonia nitrogen in water and soils that might cause the nutrients to become unavailable to active contaminant-degrading microbes)

3. Total heterotrophic bacteria:
   Plate counts
   MPN testing
4. Specific hydrocarbon degraders (for target contaminants), and Microtox® assay for microbial toxicity
5. Shake-flask biodegradation testing measuring oxygen or hydrocarbon consumption under various conditions

Once the screening data have been gathered, they should indicate if biological treatment is a viable option. If it is determined that there is a good chance for success, there are two proper courses of action available. Either (1) the remediation design is prepared from the existing data and remediation is initiated, or (2) it is determined that more data are needed and a full-blown treatability study is recommended.

In some cases, the screening data generated in the *bioassessment study* might suffice for the project at hand and the consultant may recommend moving directly to design and implementation. On the other hand, if the client (or regulatory agency) insists on more definitive estimates of cost and timing for the project at hand, more extensive testing for *treatability* will be necessary. It must be clearly understood that the rates of contaminant degradation are always site specific and should never be assumed. Field data often confirm degradation rates three to five times slower than those observed in the laboratory. Assuming rates of degradation for specific compounds has often resulted in profound professional embarrassment, sometimes followed by swift litigation. It is always best to define the expected rate of cleanup ahead of initiating the full-scale remediation, and this is best done through bioassessment or biotreatability.

BIOTREATABILITY STUDIES

If the data generated during the bioassessment screening prove to be insufficient, a more detailed testing procedure can be initiated to identify factors and parameters involved in fine tuning the bioremediation process. The objective is to identify the optimum conditions for (and to determine a general rate of) degradation of the target compounds. Treatability studies generally *do not* determine the extent of degradation or define the ultimate bioremedial end point, often referred to as "how low can you go?" That kind of information takes many weeks or months to develop and may not truly depict the actual field experience to be expected. That kind of requirement and level of knowledge can only be approached in an actual *pilot test* on a portion of the actual field site.

Biotreatability studies can employ a number of protocols. The major types will employ:

1. Microcosms
2. Respirometry
3. Bioreactors

Microcosms

A treatability study can utilize prepared *microcosms* of various sizes: (1) packed columns to simulate soil conditions at a site where water to be treated is passed through the columns, either in upflow or downflow mode; (2) VOA vials, flasks, or expensive glass apparatuses which retain all volatiles and allow studies to be conducted by sacrificing vessels; or (3) soil trays containing soils to be treated. Such vessels are referred to as microcosms, where conditions within each vessel can be varied (including nutrient and oxygen levels, microbiology, pH, etc.).

Respirometry

Another common method utilizes *respirometry.* Respirometry apparatus consists of tubes or vials that contain the soil or water to be treated. Each vial is set up as one of several microcosms (containing varied nutrient or pH regimes, etc.) and monitored for O_2 uptake or CO_2 generation. This will confirm the apparent metabolic rates between treatments. Analysis of the vial contents can confirm contaminant destruction.

Bioreactors

Possibly the most comprehensive (and costly) testing protocol for biotreatability utilizes *bioreactors.* These units consist of a biostat or biological chamber that can be filled with site media. The reactor has multiple ports for introduction of fluids, gases, and solids, and that provide for sampling of internal contents. The unit is compact and is subject to a full range of parameter manipulation and environmental simulation. They are most useful for treatment of concentrated wastes or in situations where a wide range of environmental conditions can be anticipated.

All treatability studies must include vessels which act as *untreated controls.* The purpose of the control microcosm is to determine any removal mechanisms other than biological. Many studies are designed to examine degradation under anaerobic as well as aerobic conditions. There can be *killed controls* where the microbes are poisoned deliberately using sodium azide or some other biocide in order to quantify any possible physical or chemical removal mechanisms.

Treatability studies are designed to address specific technical issues, costs, and timing associated with a project. They typically entail sampling and analysis at specific time intervals during the study to track performance of the simulated remediation systems and to test for the formation of undesirable intermediate degradation products. In fact, the analytical component of these studies can be the most significant feature and can represent the highest percentage of cost as well.

It is important to understand that not all bioremediation projects have treatability studies conducted as part of their scope of work. In many cases, physical limitations associated with a project can far outweigh some of the scientific issues, so that field pilot demonstrations become more practical and/or necessary. Treatability studies are conducted when there are specific issues related to the scientific aspects of a project that warrant quantification. These studies can span 2 to 6 months in time, depending upon the number of factors being evaluated and the particular contaminants that may be of interest. When *radiotracer compounds* are used as part of the treatability protocol (i.e., in order to track a specific degradation pathway), this virtually assures that the study will go a full 6 months in order to generate the appropriate information. Typical site sample volumes collected for treatability studies can be more than the maximum volumes described for water and soil in the screening study description. An added cost that can be substantial involves the disposal of samples and culture media (as when site materials might contain hazardous constituents).

Cursory treatability studies cost between $10,000 and $20,000. This is in addition to the cost of the screening study for a project. The overall cost range for a *full biotreatability study* is $20,000 to $100,000. This range is representative of the many different factors that can be evaluated in such a study, the large volume of samples that may need to be processed, and the role of sophisticated analytical methods in the testing. Some laboratories offer fixed fee studies, but many are still developing their capabilities in this area, making time-and-materials charges more common. Also, since there can be few standard sequences of techniques for biotreatability studies due to site specifics, most laboratories have developed their own special protocols.

CHOOSING A LABORATORY

A word of caution is in order here: many labs that are fully capable of performing meaningful analyses lack the capability and expertise to perform proper specific microbiological analyses and associated data reduction and interpretation for a bioassessment. In this regard, the site owner and/or remediation specialist who must perform the remediation (and close the site) are ultimately at the mercy of the chosen laboratory concerning the initial data for system design and monitoring program, and for proper execution of the project. Therefore, it is of paramount importance that the laboratory management and personnel possess more than just a passing acquaintance with bioassessment testing protocols and procedures. Lab selection guidelines that assess experience and certification should be followed, and the client is well advised to rest on the advice and counsel of the bioremediation specialist who is ultimately responsible for the success of the project.

As discussed above, the data resulting from bioassessment or biotreatability must be interpreted by a consultant or remediation specialist who is competent in applied bioremediation. One might inquire, "*How can competence be determined*?" In light of the authors' combined 70 years of experience in environmental remediation work, it is our opinion that a competent practitioner in this field should be able to list several verifiable sites and clients where a site closure was obtained through applied field bioremediation. Failing this, a list of ongoing projects or current clients should be provided. In our experience, a single site closure testifies to a simple familiarity with the technology. Several site closures attest to a considerable knowledge in the field. There are precious few environmental professionals who can point to more than a handful of successful field closures (outside of oil field production pit treatment, etc.). Although many researchers are quite familiar with the field sampling and laboratory protocols and may have published extensively, they simply lack the actual hands-on field bioremediation experience required for sound judgment in many cases. And notwithstanding claims to the contrary, that makes the competent field bioremediation specialist something of a "rare bird". The corollary to this is simply to check out the claims of potential remediation specialists *before you sign the contract.*

CHAPTER 5

Applied Bioremediation: An Overview

Now that we have covered the technical basics, let's review the critical information that requires understanding in order to ensure a proper application of field bioremediation. This chapter covers the application of essential elements and engineering principles to accomplish an actual field bioremediation and site closure.

Common questions asked by waste generators these days are

"Can bioremediation work for me?"
"How does it work?"
"How fast will it work and how much will it cost?"

Generally, the answers (from a qualified bioremediation specialist) will sound something like "If your site contains crude oil, fuels, or other organics (lubricants, chemicals, etc.) in acceptable concentrations, there is a chance that microorganisms will degrade them. The cost and timing will depend upon your site's characteristics."

THE BIOREMEDIATION APPROACH

Bioremediation can often be significantly less expensive (1/3 to 1/2 the cost) when compared to other alternatives. In the case of surface treatment of soils, it will usually take more time to complete when compared to excavation and landfill disposal. However, when applied *in situ* for groundwater treatment, it can reduce the time severalfold compared to conventional pump-and-treat technologies, which can take 15 to 30 years without completing the job.[1] Although bioremediation of metals is just an emerging technology at the present time, there is a chance it can restore certain environments.

Bioremediation is the application of the normal metabolic processes of microorganisms to alter the chemical structure of organic materials in solid, liquid, or vapor states in order to render the compounds less toxic to humans and the surrounding environment. It is also the use of microbial metabolic processes to change the valence state (or redox potential) of metals for detoxification, or to render them less soluble, and therefore unavailable for transport through the environment. This is accomplished by manipulating the environment of the site in ways that encourage multiplication and fast metabolism of the microorganisms which can degrade or detoxify the unwanted contaminants (see Chapter 3).

Bioremediation is most often employed to change the molecular structure of organic compounds (those with a carbon-based "backbone"), thereby degrading toxic or potentially toxic organic chemicals into "harmless" compounds. This is a *destructive process* as defined by EPA, and thus is a preferred method of dealing with hazardous materials, since the hazard is permanently removed by the change in chemical structure. The most common end products of aerobic bioremediation are carbon dioxide, water, chloride ions (where chlorinated hydrocarbons are degraded), and partially digested "daughter compounds" which are sometimes further degraded. Completing the conversion of organic carbon to inorganic carbon (in the form of carbon dioxide) is called *biomineralization* (see Chapter 3). If the reaction is incomplete and produces some of the "daughter compounds," the process is said to have been merely a *biotransformation.*

This conversion is accomplished by microbes via their enzyme metabolic pathway systems, which attack the organic molecules in solution. Since all enzyme reactions occur in aqueous solution, this means that water must be present in sufficient concentrations during the remediation. Water is essential whether the microbes can pass organic substrates through their cell membranes for *intra-cellular digestion*, or whether they must pass enzymes to the outside for *extracellular digestion.* The enzymes capture clusters or individual carbon atoms from the organic contaminant molecules to use as a carbon source for respiration (energy production) and to build new cells (biomass production).

Another requirement of the microbes to build cells is the *macronutrients*, primarily the proper forms of nitrogen and phosphorus, since these elements are major components of all cells (see Chapter 2). These nutrients often must be added to the matrix (soil or water) to be treated in order to encourage the microbial population to grow and carry out the remediation. Once an initial supply of these nutrients is established, the *lysing* of dead cells can replenish the nutrient supply if the microbial population is not growing, so the remediation can proceed at some rate even if the nutrients cannot or will not be further amended. Other elements which may be required in some concentration are sulfur, potassium, and sodium, and other metals such as copper, magnesium, zinc, and iron. These are typically called *micronutrients* because the required supply is very small (not much is needed). Some metals can be toxic to microbes in high concentrations because they can interfere with the metabolic reactions within the cells. Extra care must be taken if a site contains high concentrations of metal contamination.

Bioremediation may not be effective in treating metals in soils, sludges, or solution. The ultimate elemental structure of a metal atom cannot be altered microbially except for changes in the valence state (electronic charge in solution). Toxicity is often associated with valence state, and solubility to some extent.

Some bioremediation practitioners claim to be able to remediate metals. Actually, the processes employed either trap or precipitate the metal on the external cell membrane or inside the cell, forming a complex microbial structure for future harvesting. This is a form of *bioaccumulation*. When a microbial process changes the valence state of a metal to a less toxic form, or alters solubility, the process is a form of *biotransformation*. This new approach to field biorestoration will become a reality with continued research.

INTRINSIC BIOREMEDIATION

Intrinsic bioremediation, also called natural attenuation or passive bioremediation, arises from the observation that in nature, microbes will degrade organic materials at some rate, even if it is not very rapid. Since the rate of movement of many organic materials in subsurface environments is not very rapid, the rate of degradation at some sites can be demonstrated to be at least as rapid as the movement of the organic materials in the subsurface. Hence, as long as there are no pathways of exposure (means of contact) to the contaminant plume, an argument can be made to leave everything alone (while continuing to monitor the plume) and the problem will eventually go away by itself. This position is valid as long as the reasoning is based on site-specific conditions, the exposure pathways can be effectively kept closed, and some form of monitoring is performed to demonstrate that the underlying assumptions on plume dispersion and movement are true.

In general, intrinsic bioremediation has been successfully used for aquifer restoration at sites where the sources for recontamination have been removed from the vadose zone (the soil zone extending between the ground surface and the water table), there is no *separate-phase hydrocarbon* present on the water surface, and the compounds present are all biodegradable. It has also been used at sites with very "tight" soil structures, where the rate of plume dispersion is essentially zero.

The amount of investigative work required to successfully apply intrinsic bioremediation may be greater than those efforts described below in a normal investigation, because the delineation work and the fate and transport calculations required to demonstrate that there is no risk to human health and the environment are likely to be more stringent than those required for a "typical" bioremediation project. The opportunity, of course, is to eliminate the cost of remediation altogether, and complete the cleanup by simply monitoring the contaminant plume until it is naturally reduced to a size or concentration previously agreed upon.

SELECTION GUIDE FOR APPROPRIATE BIOTECHNOLOGY

Considering all the possible sites and circumstances that are amenable to biotreatment technology, it is helpful to try and determine early-on the type of treatment for a particular site that will deliver the most effective treatment from the standpoint of (1) cost, (2) timing, (3) cleanup efficiency, and (4) regulatory familiarity. These factors are commonly written into many requests for proposals (RFPs) and their significance to a specific project is often weighted in terms of the decision criteria being applied to each factor.

In choosing an appropriate biotechnology for remediation of a specific waste problem and site conditions, it is useful to proceed in an orderly fashion toward a justifiable selection on the basis of good information. It should be the goal of the waste treatment professional to make (and document) every decision based on a sound evaluation of the merits and limitations of those technologies that have application. With this aim in mind, what follows is a discussion of probable decision factors and a decision-matrix arrangement of considerations for choosing an effective biological treatment alternative (see Table 5-1).

Table 5-1 Biotreatment Decision Matrix

	Land treatment	*In situ* treatment	Bioxidation treatment	Bioreactor treatment
Soils	X	X		X
Sludges	X			X
Ww streams			X	X
Groundwater		X	X	X
Lagoons		X	X	X

It is important to note that many other factors can play a role in the decision to employ biotechnology in a given site remediation. Some of these may be quite apparent, while others, though subtle, may be equally important. It is rare that a single selection criterion will dictate the specific bioremediation technology that is most appropriate. The following considerations are important and have been used often by the authors to determine the appropriate technology for use at many sites. Selection should be made for the approach(es) that:

- Most effectively degrades the target contaminants
- Best addresses remediation of the matrix or phase in which the contamination is present (dissolved, adsorbed, or both)
- Has proven to be the most effective in transporting nutrients and oxygen (or other electron acceptors) to the bacteria at the point where contaminants are present
- Will make use of as many existing site conditions, utilities, and labor materials as possible
- Has generated positive case histories for this type of matrix and contaminant
- Is least disruptive to the facility (whether it is active or there is other environmental work ongoing at an inactive facility)
- Can be most readily permitted by the regulatory agencies involved, or that is deemed acceptable by those regulators
- Can be most rapidly deployed and started up
- Best suits the site layout or is most mobile for use at other locations within the site
- WILL CLOSE THE SITE (reach target levels, pass TCLP assay, or meet the guidelines of a risk assessment)
- Will allow salvage of the capital equipment once the project is closed
- Will offer the most flexibility in system operation through training the client's employees as operators, or that allows procurement of goods and services at direct cost
- Will be most inconspicuous at the site
- Best protects human health and the environment

- Will serve as a demonstration of the approach for use at other locations or sites where the client is responsible for cleanup
- Will most readily complement other technologies to be utilized at the site
- Will not require excessive treatment of offgas, sludges, or other residuals
- Will best reduce or eliminate client liability

Soils and Sludges

Soils can be treated *in situ* or excavated for bioreactor or surface land treatment. Generally, if the contamination consists of organics or petroleum hydrocarbons and time is not a constraint, the most cost-effective method is *in situ* bioremediation with bioventing as an additional option (see Chapter 11). If the quantity of soil requiring remediation is only several hundred cubic yards, the soil may be excavated for more timely bioreactor or land treatment. However, if time is short, conventional hauling and landfilling (with the specter of continuing liability) or asphalt blending may be considered. It is rarely necessary to incinerate soils; from a cost standpoint, it is unthinkable, except for small quantities and perhaps for certain recalcitrant compounds or toxics. Low temperature desorption and detoxification is a new technology that avoids some of the hazards of material handling, the liabilities incumbent with landfilling, and stack gas emissions associated with incineration, and in the case of some waste, can be operated without the necessity for EPA permitting.[2]

Sludges containing organic chemicals and hydrocarbons can be land treated or can undergo bioreactor processing. When faced with possible landfilling or incineration, the nominal cost of a biofeasibility study should be pursued in the interest of enormous potential for cost savings compared against incineration, if time is not a critical constraint. The least that can be accomplished through biotreatment is a substantial percentage reduction in the quantity of sludge requiring ultimate disposal.

Liquids and Aqueous Solutions

Surface waters and wastewaters containing a wide variety of soluble organics and inorganics are especially amenable to biological treatment. In the case of impoundments, *in situ* bioremediation can be easily accomplished (see Chapter 10). This technology can often suffice as a means of closure or it can substantially reduce the bulk of material to be handled by other means. Biotreatment of surface waters and wastewaters can be applied *in situ* or by means of a treatment train employing any of the technologies described in Chapters 6, 8, 9, and 10.

Groundwater contaminated by organic solvents or fuel hydrocarbons should ALWAYS be assessed in terms of *in situ* bioremediation prior to implementation of any other alternative. For example, if it is determined that a standard pump-and-treat method with surface treatment will be employed, simultaneous nutrient injection of a portion or all of the produced water will (1) speed the time to completion of the project, (2) ensure a clean closure, and (3) conserve precious groundwater resources. In addition, surface biotreatment of extracted groundwater could prove to be more cost effective than other surface treatment alternatives. Combinations of surface and *in situ* biotreatment simultaneously employed are becoming more popular (see Chapter 9).

PROJECT ORGANIZATION

The successful practice of bioremediation of wastes and contaminated materials requires the completion of a sequence of steps, each of which must be executed successfully to realize the objective of spending the time and money: *a site that has been declared "clean" by the appropriate authorities*. This is the ultimate objective, whether the work is done by contracted consultants, remediation firms, government contractors, or the actual site owner. Pressure and demands from regulatory agencies in any project may range from none at all, to a consent order, to the pursuit of a **potentially responsible party** (PRP) under SuperFund. In any case, the "product" is still defined as a clean bill of regulatory health. Everything undertaken in the course of a project should be directly related to this goal.

The approach to most bioremediation projects must include the following essential steps:

1. Site characterization
2. Initial feasibility testing
3. Detailed biotreatability testing, pilot testing, and design (including hydrogeology)
4. Procurement, installation, and start-up
5. Process monitoring and operation
6. Final sampling and closure
7. Reporting and management

Some of these steps are consecutive to the previous step, some steps overlap, and others (i.e., reporting and management) continue throughout the project. Each of these steps is discussed in some detail below. However, this discussion is not intended to produce a set *recipe* for selecting a remediation approach, because there are few *typical* sites where only one process would or could apply. These are, nonetheless, presented as useful generalizations that can provide a framework on which to build the site details and develop an approach which can produce the desired result. There are, however, several more common scenarios for treatment of released organics such as leaking underground storage tanks (LUSTs) at gasoline stations (gasoline, diesel, used oil), dry cleaners (trichloroethylene, carbon tetrachloride, etc.), parts cleaning facilities (various solvents), and storage terminals (petroleum fuels and liquid products), where releases of similar types of materials often require cleanup.

SITE CHARACTERIZATION

The objective of the characterization phase is to answer three key questions:

1. *What* is the contamination (including the materials which are high in toxicity) and what are the *concentrations* of those materials?
2. What is the *soil texture*, and can materials be moved through it using air or water as carriers?
3. Is the contamination above, below, or in the *water table*, and what is the *degree* of contamination of these horizons?

These questions require that the site chemistry, geology, and hydrogeology be initially characterized in order to determine whether bioremediation appears to be a viable option. If these conditions are favorable, then further work can be done as the project progresses to get a sharper picture of these site parameters and to develop an effective bioremediation design.

The site characterization often begins by searching federal, state, and/or local agency records concerning the site and its history. This activity may either begin as (1) a "due diligence" property transfer investigation as part of a real estate transaction, (2) a notice of violation (NOV) for an operating permit at a facility, or (3) a voluntary action because the owner knows the site is contaminated as a result of previous or current activities conducted there. Typical questions to be asked include:

- Have any spills or leaks been reported?
- What chemicals were stored, used, or handled at the site?
- What agencies have responded to incidents at the site?
- Are there or were there any underground storage tanks at the site? What was stored in them?

Most property transfer assessments (PTAs) in the U.S. are now conducted under ASTM Standard E1527-94, *Standard Practice for Environmental Site Assessments: Phase I Environmental Assessment Process* (June 1994), which requires that many site-specific questions be answered in order to consider the assessment complete. Appropriate maps of the site and the local geology may also be obtained during these searches. There are several standard data base search companies who will conduct at least portions of these searches electronically.

The next part of a site characterization is often a *site walk* to review the locations of potential trouble spots. This is generally done as part of a PTA, which visit is not specifically searching for evidence that bioremediation will work at this site. **BRING YOUR CAMERA!** If the owner will permit it, use a video recorder with a running commentary of observations; this could protect both you and the owner in the future by providing a factual description of what was there at the time of the site walk. This activity is the first opportunity to estimate how extensive the potential contamination is, and what may be required to clean the site. It is also the time to determine where to make borings for initial samples of soil and groundwater (if appropriate) in order to begin outlining the extent of the contamination or to prove that no contamination exists.

After, the site walk (or the PTA) indicates that sampling is necessary to determine the extent of environmental impact, the site map is marked, and a detailed sampling plan is developed, often for review and approval by the appropriate regulatory agencies. You will need to determine which agency has site responsibility for each particular owner; this is highly dependent on the state in which the site exists. If the site history and the site walk indicate that bioremediation may be a viable option for cleaning the site, the sampling plan should include borings for soil and water from the locations that you expect to show contamination. Samples from these borings will be needed to begin the *bioassessment screening* step and *treatability studies*, if needed, as described in the next section (and in considerable detail in

Chapter 4). Obviously, the sample locations are not usually known in advance with any high degree of accuracy or certainty. Therefore, the samples for feasibility testing often are taken during a second sampling round to delineate the extent of contamination found as a result of the first sampling event.

The purpose of the sampling exercise is to determine, to the maximum extent possible, the nature of (and the horizontal and vertical extent of) the contamination at the site. Initial sampling locations are chosen on the basis of physical evidence, such as stained or disturbed soil, odor or other vapor detection, and site history. This sampling exercise often takes more than one sampling event to complete. Because the sampling results are returned from the testing laboratory a matter of weeks after the initial sampling was conducted, contaminant concentration trends are not available while the sampling is conducted. Broad indications can be gleaned by conducting soil vapor surveys (if the contamination is relatively shallow), by using *direct-push* methods to obtain soil or water samples for field analysis by *flame ionization–gas chromatography* (GC) or other field assay methods such as *immunoassay* test kits, or by monitoring the boreholes and samples using a *photoionization detector* (PID). This is often done for health and safety reasons. While these methods may not be considered strictly quantitative, they give at least some idea of the presence and extent of volatile organic contaminants. They will not detect heavy organic species or metals. For these compounds, there is no effective substitute for sampling and laboratory analysis.

One general note worth mentioning here involves suspected organic contamination in the vicinity of the water table. In this case, at least one sample should be taken from the top of the *saturated zone* for analysis. Many sites containing organics will be contaminated at this depth, and the contaminant concentration will be important both for estimating the total quantity of material to be treated and, hence, the treatment time required.

As the field notes and the results of the laboratory analyses become available, several items should be considered in determining whether bioremediation is a viable option:

1. As described briefly above, the presence of heavy metals substantially above regulatory threshold values indicates three potential difficulties: bioremediation will not remove large quantities of these materials; however, microbial processes may alter their solubility; or they may be inhibitory or toxic to microbial action. Bioremediation may be appropriate only as a partial solution and additional technology may have to be employed to effectively treat the metals, or the remediation may employ bioremediation for both organics and metals contamination applied in a sequencing fashion.
2. The presence of *polycyclic aromatic hydrocarbons* (PAHs) with more than five rings indicates that the contamination will be degraded slowly, if at all, using indigenous bacterial populations (i.e., bacteria already living in the soil and water at the site). For these materials, some form of excavation and/or blending (composting) may be considered, or another type of treatment may be more cost effective.
3. Highly chlorinated hydrocarbons (three chlorine atoms per molecule and higher) present several difficulties in aerobic degradation, particularly in the absence of

nonhalogenated hydrocarbon materials. The most common compounds in this class are trichloroethene (TCE), tetrachloroethene (perchloroethylene, or PCE), and polychlorinated biphenyls (PCBs). In the absence of a cometabolite, these highly chlorinated species are not normally degraded by indigenous aerobic populations. This is also true for halogenated pesticides. For a more detailed discussion of this issue, see Chapter 3.

4. During the process of drilling soil borings, a competent geologist should *log* the borings using a generally accepted reference system to classify the soils encountered. The industry, in general, appears to be standardizing on the *Unified Soil Classification System* (USCS) for most soil borings performed. These boring logs provide an indication of the soil type that exists at the subsurface depths or *horizons* at which contamination has been encountered and, in turn, provide the means whereby an estimate can be made of the soil's ability to transport air, water, nutrients, contaminants, and by-products into and out of the contaminated zone.

Any or all of these factors, if found to be incompatible, can potentially rule out bioremediation as an effective treatment method. If under these conditions bioremediation is excluded, this determination can be made with a minimum of financial expenditure. While it may not be immediately gratifying to do this (particularly if bioremediation is the only technology being considered), in the long run it is advantageous. **BIOREMEDIATION DOES NOT WORK AT EVERY SITE, IS NOT NECESSARILY THE LEAST EXPENSIVE METHOD AT EVERY SITE, MAY NOT MEET THE REMEDIATION TIMETABLE AT EVERY SITE, AND MAY NOT ACHIEVE THE NECESSARY CLEANUP STANDARDS AT EVERY SITE.** Fortunately, though, for many sites, the technology does accomplish all of these things.

After site characterization, you should have a good idea of what contamination affects the site, where it is located, and whether or not the site groundwater is involved. The next step is to determine, as inexpensively as possible, whether other factors are involved that would prevent the successful application of bioremediation. This can be accomplished through a *biotreatability study.*

The discussions below must include the most complicated situation: contaminants have been found in both soil and groundwater (or sludge and water above it). Much of the discussion below deals with a determination of the flow behavior of the site groundwater, but this situation only applies when *in situ bioremediation* below the water table is being considered for a site. When other technical approaches are being considered, this information may not be needed. If the presentation seems overly favorable toward *in situ* treatment, that is the authors' choice for many reasons. Specific methods for each technical approach are found in subsequent chapters.

INITIAL BIOASSESSMENT TESTING

The details of a feasibility test vary depending on the preferences of the company or individual, the equipment available at the laboratory, and the degree of testing the site owner is willing to fund. The cost range is from as little as \$2000 (for a cursory nutrient study) to over \$50,000 for a detailed *biotreatability study* which includes biostimulation testing, and geochemical transport tests. The initial testing for a major site typically will cost between $5000 and $20,000 as a compromise of the above factors.

The essential components of the testing are heavily affected by the need to test for water contamination as well as soil contamination. If the groundwater is contaminated, the costs rise dramatically compared to soil testing alone.

A typical testing protocol for the initial *bioassessment screening study* includes the following:

- Determination of soil, water, and/or sludge *nutrients*: ammonia nitrogen and orthophosphate concentrations.
- Soil, water, and/or sludge pH and percent *moisture* (for saturated as well as unsaturated regimes, as appropriate).
- Microbial characterization of soil, water, and/or sludge in the form of *plate counts* for total heterotrophic (organic carbon-consuming) bacteria, most probable number (MPN) tests for specific contaminant degrader organisms, and possibly a nutrient stimulation and plate count to determine the effect of adding nutrients to the soil/sludge/water matrix. A variation on this characterization is the use of a Microtox® bioassay to determine the *relative toxicity* of the soil/sludge/water matrix to a sensitive strain of bacteria.
- If the remediation is to proceed in groundwater, a *nutrient transport test* will be needed for orthophosphate and ammonium ion in solution through the aquifer soil. In addition, a test for the ability of the native groundwater to hold the nutrient blend in solution for a specified period of time (usually 24 hr) will be needed. Water containing calcium in any significant concentration will cause precipitation of orthophosphates and may require a change in nutrient formulation to prevent the soil formation from plugging or cementing at least part of the available pore spaces.
- If an oxygen source is contemplated for addition to the circulated water, a *stability test* at various concentrations of the O_2 source is needed to determine any transport and decomposition difficulties. This may be a hydrogen peroxide stability test in the site soil and groundwater, or simply the ability of the site soil and water to disperse air or bottled oxygen directly injected without producing by-product reactions. A variation on this test is a measurement of *biological oxygen demand* (BOD) and *chemical oxygen demand* (COD) on the site water, to determine whether there will be significant competition for the oxygen from other sources (e.g., iron oxidation).

The screening tests above are intended to define any of the *knockout factors* that might rule out bioremediation as an effective treatment alternative beyond those discussed in the site investigation (they also represent an increased level of detail). Some of the problem areas that can be identified as a result of the *bioassessment screening study* are

- Lack of a suitable indigenous bacterial population (attributable to a toxicity problem, or other inability of the native population to degrade the pollutants)
- Precipitation of essential nutrients by soil components (especially calcium precipitation of orthophosphate)

- Soil interactions with the nutrients to retard their transport through the soil
- Unsuitable environment initially as indicated by pH or Eh problems (this can also be indicated by low bacterial counts)
- Stability problems when introducing oxygen-bearing compounds below the water table. (This will limit the ability of the groundwater to effectively carry oxygen from hydrogen peroxide, bottled oxygen, or other sources through the contaminated zone.)

DETAILED STUDIES, PILOT TESTING, AND PROCESS DESIGN

At this point, assuming that the *bioassessment screening study* has not identified any knockout factors which would eliminate the use of bioremediation, a detailed series of site-specific tests, (which may include a biotreatability study) is required to develop a design for remediation. Again, the tests required are determined by the need to work with soil, water, sludge, or some combination.

The discussion below is based on the need to treat soil and water; if only one medium is involved, only that set of tests needs to be performed. The discussion also assumes that the remediation will be conducted in aerobic mode, so that oxygen will be the terminal electron acceptor.

1. A **biotreatability** study may be needed to determine the relative rate of degradation and removal of the target contaminants. The extent of testing will depend on the equipment available and the target compounds of interest. Some of the options include:
   - The use of a *respirometer* to measure oxygen uptake or carbon dioxide generation by the bacteria in a closed system, as an indicator of the degradation rates of individual contaminants and the oxygen supply rates required.
   - A *microcosm study* to investigate the effect of a series of nutrient doses to determine the optimum addition rate and frequency (usually accompanied by an abiotic control, poisoned with a bactericide to measure nonbiological effects of the treatment protocol).
   - Regular sampling and analysis for the target contaminant concentrations as the remediation proceeds, and in some cases, the predicted degradation by-products. Because they are quite expensive, these analytical procedures will tend to affect the cost of the design phase more dramatically than any other component. For example, a full organic pollutant scan of every sample during the degradation study is quite expensive at a cost of $800 per sample; 30 such samples would not be considered excessive for this type of work. Custom analyses for specialty compounds would also add to the cost. You may want to consider frequent analysis using a broad-based parameter such as *total organic carbon* (TOC), *total petroleum hydrocarbons* (TPH), or *oil and grease*, all of which are relatively inexpensive. Less frequent use (biweekly or weekly) of analyses for all the target compounds and daughter products to provide the necessary detail as to degradation rates can minimize overall analytical expense of the project.
2. **Aquifer mechanics** tests for vertical and horizontal hydrological control of the groundwater in the contaminated zone are required to determine the rate at which water can be moved through the area. These tests may also involve several degrees of complexity and completeness. Some of these options include:

- A simple *percolation test* such as that used for a septic system provides information on vertical permeability near the surface and/or above the water table.
- A *slug test*, where water is added to a well in a "slug" (single large dose) and the water level in the well is measured over time as the water enters the formation, provides a rough estimate of the vertical and horizontal permeability of the soil and its ability to transmit water through the aquifer.
- A simple *pump test* (where a well is emptied of water and the level is measured over time as the well recharges) provides the same type of information as a slug test.
- A *static ring test*, which is a variant of the percolation test using specially designed equipment, provides a more precise absorption rate than a percolation test.
- An *extended pump test* in which a well is pumped for an extended period (ranging from 24 hr to 6 months) determines the longer-term sustainable pumping rate and its effect on the water table in the vicinity of the well.

These tests provide data on the *aquifer transmissivity*, *storage coefficient*, and other characteristic flow parameters. For *in situ* bioremediation, the essential data include the aquifer *yield* measured in gallons per minute of water, and the groundwater flow *velocity* measured in feet per day. This is not the same as the aquifer transmissivity, usually measured in cm/sec, and the conversion from one measure to the other is not just a simple change in units. These tests determine (within an order of magnitude) a critical factor: an initial estimate of the time that the remediation will require. The degradation rate, coupled with a knowledge of the extent of the contamination and the rate at which materials can be transported through the contaminated zone (from the site characterization phase), permits the design of an extraction/injection system that will move water and nutrients through the contaminated zone at the appropriate rate to complete the cleanup.

For *in situ* groundwater treatment, the system must be designed to:

- Remove contaminated water to the surface
- Remove the contaminants from the extracted water at the surface
- Add the necessary nutrients and oxygen to promote aerobic microbial activity in the contaminated zone
- Reinject a part of the extracted water (along with the added nutrients and oxygen) into the contaminated zone

This technology is sometimes called the Raymond process[3-5] after the name of its inventor. The basic design principles for these requirements are as follows:

1. Groundwater recovery wells are usually placed at the perimeter of the contaminated zone at the lowest point of the *hydraulic gradient* (downgradient), often where the ground level is the lowest. It should be taken into consideration that the action of recovery wells or trenches can further lower the water table in the immediate area, and may pull water supplies from the area immediately beyond the recovery zone.
2. The injection gallery, trench, or wells should be placed at the perimeter of the contaminated zone at the highest point of the groundwater gradient (upgradient), often at the highest elevation end of the contaminated zone. This action will raise the water table in the immediate area (commonly termed *mounding*). Water from

excessive reinjection can also tend to push contaminated water to the side of its intended pathway and cause it to escape the capture radius of the extraction well. It is at least very difficult, and may be impossible, to reinject all of the recovered groundwater. The mounding problems associated with complete reinjection, and the potential for plugging or cementing the reinjection wells with calcium precipitates, make for a potentially difficult situation if alternative outlets for the treated water do not exist. Some sites where 100% reinjection has been tried have been forced to operate at well below the design flowrates until another permitted outlet for the water could be developed, either because the wells plugged off and required reconditioning, or because natural rainfall created flooding at the surface.

3. The water brought to the surface usually requires some form of treatment to remove any contaminants which would prevent the water from meeting reinjection standards. This treatment may take the form of a *bioreactor*, or it could include an *air stripper* for volatile organic hydrocarbons, *ion exchange* or *precipitation* for metals, *carbon adsorption* for heavier hydrocarbons, pesticides, or other adsorbable materials, or other forms of treatment depending on the nature of the contamination.
4. Nutrients and an oxygen source are added directly to the water prior to reinjection. Often, nutrients are added in batches (pulse fed) while the oxygen source is added continuously. The concentrations to be added and the frequency of nutrient addition should be determined from the *biotreatability* testing, as described above.
5. The remediation system should be designed so that the transport time from injection to recovery will be less than 3 months if possible, and less than 1 month if the installation costs will permit it. A factor which greatly influences the design transport time is the installation cost for the remediation system. Spacing wells farther apart, although tending to increase the remediation time and residence time in the contaminated zone, may still be more cost effective than installing a greater number of wells, trenches, galleries, or similar systems. A more complete analysis of the site hydrology can help to determine the most cost-effective approach to well spacing and design residence time for circulated groundwater.
6. All of the geological data collected for the site should be incorporated into a *hydrogeological model* in order to determine the best arrangement of injection and recovery hardware. There are several rewards for diligently modeling the hydrogeological factors at this stage:
   - The injection and recovery systems can be placed where they will be most effective and thereby minimize the installation costs.
   - Modeling provides a basis for calibrating the site flows, so that the distribution of materials can be analyzed and understood when the remediation process begins.
   - A troubleshooting tool for system performance is readily available so that unexpected contaminant or nutrient concentrations or water level measurements can be interpreted.
   - *Hydrogeological modeling* provides data demonstrating that the contaminants are contained within the design capture zone, thus meeting the requirements of regulatory agencies. Many regulatory agencies require evidence that the flows are going where you say they are (while the model itself will not serve as proof to a regulatory agency, its calibration to the site will serve as the best evidence available).
7. The site flow characteristics, coupled with the contaminant removal rate (as a result of biological action), predict a remediation time requirement, usually measured in months. More rapid flows through more permeable formations will tend to allow for a faster treatment rate.

As noted above, when the site groundwater is not affected, this entire process becomes much simpler and less expensive. Steps 1, 2, 4, and 5 of the preceding discussion can be eliminated when water is not used as the transport medium, as in composting, land treatment, biovaults, bioventing, biofilters, and many bioreactor applications. In these cases, nutrient loadings and degradation rates are used to provide cost and remediation time estimates.

If time permits, it is sometimes desirable to conduct **pilot testing** of the biological treatment system on-site. Typically, a section of the site is selected and the actual system (or other similar test apparatus) is run for a period of 3 weeks to 6 months. Various modifications can be made during the pilot test period to determine the best operating conditions for the system. Employment of a stringent analytical protocol is highly beneficial during the pilot test in order to thoroughly document treatment efficiency.

Pilot tests are most often conducted when the owners or regulators are inexperienced in dealing with bioremediation technology and require a *"confidence boost."* Pilot tests are also conducted when there are technical issues which need to be resolved in large projects before proceeding to full-scale operation. The resolution of these issues can greatly affect the total dollar amount spent by the owner.

At this point, the system design is sufficiently complete to provide predictions about a cost estimate and remediation time requirement with some degree of confidence. Some notes of caution are

1. No site is absolutely homogeneous geologically, although some sites come close. A prediction is only an estimate, the accuracy of which is based on the geological stratum tested.
2. Temperature has a considerable effect on the metabolic rates of microorganisms. Therefore, temperature is a critical factor in determining a realistic remediation time requirement. Since we cannot usually control the climate, the remediation is dependent for the most part on seasonal and ambient temperatures. Groundwater does not change temperature as much as shallow soil; however, there are seasonal variations in water temperature which can affect the degradation rate.
3. The test work described above is usually conducted for an estimate of *average* concentrations of contaminants. A prime consideration for predicting a remediation time requirement should include the fact that *degradation rates* may decline over the course of the remediation process. The initial degradation rate may be higher for materials of low toxicity at higher than average concentrations. At substantially lower contaminant concentrations toward the end of the remediation, the rate may slow considerably as the microbes run out of food.

All of these factors should be considered sufficient cause to quote cost estimates on a basis other than lump-sum. Lump-sum bids are generally not practical or reliable, as the uncertainties discussed above demonstrate. A time-and-materials estimate is a more practical bidding method because the uncertainties and risk are shared. An alternative is to bid the design, installation, and start-up of the remediation system as a lump sum, and operate the system for a fixed monthly fee based on some predetermined level of effort.

PROCUREMENT, INSTALLATION, AND START-UP

Now that you know what your system has to do, it's time to go out and buy it. The equipment can be purchased as a series of components for later assembly, or as a pre-assembled *turnkey system*. There are at least several vendors for each type of equipment required. In general, reputable pump, tank, and mixing equipment vendors are willing to provide advice as well as prices. At this point, if you are not a *project manager*, it is advisable to find and hire one. This phase of the project presents the greatest opportunity to waste a considerable amount of money and time with little, if anything, to show for it at the end. Murphy's Law gets quoted frequently at this stage, and it is usually prudent to estimate the time the installation and start-up would take under perfect conditions, double that figure, and then add a week or two for extra problems. Regulatory permits have to have been received in order to start up the system, although procurement can begin for long lead-time items before the permit is received if the agency has pre-approved portions of the design. This aspect of a project is not for the faint hearted, and planning well in advance of the projected start-up date can save a good deal of lost sleep.

The details of a site will dictate the particular equipment requirements and the installation and start-up sequence. In general, the following areas need specific, detailed consideration for the start-up to work well:

1. Is there power to the site, and is the power of the right voltage and phase to meet the equipment needs? Do you need any other utilities? Are there requirements for explosion-proof motors and intrinsically safe control systems?
2. Is the operating equipment in a secure area, or do you need fencing or security guards? Are weather conditions likely to create a need for special shelter requirements for the equipment? Do you need heating or air conditioning for electronic monitoring equipment?
3. Is the process control system designed to shut off automatically if the system is in danger? Who will know if this happens, and how? How does the process behave during a power failure? Can you rig for this contingency?
4. Does the site *health and safety plan* include a local project contact, and police, fire, ambulance, and hospital phone numbers? Do these people know what you are working with and how to respond to a site problem?
5. Have you and the regulatory agencies agreed to a cleanup standard for soil and water at which point the work can cease? Have you got it in writing?

Someone has to put all this equipment together and get it to run. This is usually the task of a *project engineer* familiar with the requirements of a remediation system. If you are not a project engineer, you will need to find one unless you are exceptionally talented with equipment and mechanics. You will also need to train the people who will run the system and conduct the monitoring in how the equipment behaves. All of these factors explain why many projects are handled on a turnkey basis by mechanical contractors. The last item the contractor turns over to the operators should be a detailed operating manual for the unit.

PROCESS MONITORING AND OPERATION

At the point where the equipment is running properly (any necessary model calibrations have been conducted and the heady excitement of start-up is over), the tedium of process monitoring sets in. In general, some operations occur more than once per month, and more frequently for systems involving water transport. The general activities consist of checking the soil (and water if appropriate) at regular intervals for the following:

1. Microbial health as measured by population counts or relative toxicity.
2. Soil nutrient conditions as determined by pH, ammonia nitrogen, and orthophosphate content.
3. Water conditions as determined by dissolved oxygen, pH, conductivity, oxidation/reduction potential, and dissolved ammonia nitrogen and o-phosphate content. As appropriate, dissolved iron, manganese, calcium, and other elements may be monitored if they could affect the progress of remediation.
4. The contaminant concentrations will normally be monitored on a regular schedule, with frequencies ranging from monthly to quarterly. In the early stages of monitoring, the frequency may be higher than this if accurate degradation rate data from the field are required.
5. Nutrients and oxygen will need to be added on a schedule initially determined by the feasibility testing and made a part of *site work plan*, but increasingly determined by the data from the most recent sampling event. Adjustments to pH may also be required based on the data obtained above.

Special cases abound, and the generic site, even from a monitoring point of view, is something of a myth. These guidelines have been used in the past and have accomplished the stated aims of reducing the contaminant loadings to acceptable levels and obtaining closure of the site.

FINAL SAMPLING AND CLOSURE

When the process monitoring indicates that the remediation target concentrations (agreed upon in the work plan) have been reached, a final sampling event is planned and implemented. The sampling intensity and the types of analysis may be more detailed for this event, because the objective is to determine the final concentrations of all the target contaminants across the site. Often the analyses performed in conjunction with process monitoring are broad parameters such as oil and grease, biological oxygen demand, chemical oxygen demand, total organic carbon, or total petroleum hydrocarbons; the desired output from these is the general downward trend during remediation of the contaminants. The analyses performed for a final sampling event must target the contaminants listed in the work plan and might typically also include *priority pollutant* volatile and semivolatile compounds plus a library search of additional compounds, the possibility of a metals analysis, and a pesticide/PCB scan. The objective here is to verify that all of the listed target contaminants have been reduced to acceptable levels.

Here the trade-off between economics in analysis and the need for detailed fate and transport data to demonstrate a detailed degradation pathway can be the subject of much discussion. The owner wants the site cleaned for the most reasonable cost, and the regulatory community is driven toward the most detailed analyses to provide the best possible data base from which to determine the end point. The amount and type of sampling and analysis performed are previously negotiated items for each site and should be an integral part of the approved work plan.

After all this negotiation, the final round of samples is taken and analyzed for everything appropriate. When the results come back, a final report is submitted to the regulatory agency summarizing the data and requesting a final review and cessation of operations. Usually, the response from the agency (when it comes) is an agreement to suspend current operations. Very rarely will a site be declared "closed." A "*temporary leave of responsibility*" is a more common result. The regulatory community is usually in a position only to comment on past activities and the present state of the site; it is not in a position to judge the future. If any operations continue at the site, the agencies need to maintain the ability to enforce the regulations. Thus the word "closure" is not normally employed by the regulatory agency.

REPORTING AND MANAGEMENT

These functions are required throughout the project. The regular reporting of the status of the project to the owner, the regulatory agency, and any other interested parties as required is an important part of the execution of the project. The objective of these reports is to inform the key participants as to the current state of work, not to provide weekend reading. The best reports contain the basic information and an interpretation of its meaning, without expecting everyone on the project team to have the same technical expertise. They can be short and concise.

The same may be said of project management. There is an art to spending enough time to successfully reach the end of the project without calamity, and without devoting one's career to a single project. How much time and energy is enough is often best determined by hindsight. The administration of a project requires the time of people — whom the owner may have never met — to keep track of the business aspects of the work. These aspects of a project are the subject of many books and seminars in themselves; it suffices here to say that they cannot be ignored or skimped.

Hopefully this overview has provided some insight into the mechanics of executing a field bioremediation project. The following chapters will discuss many of these aspects in considerably more detail, and provide some anecdotal examples of the authors' experience in applying these technologies.

REFERENCES

1. HMCRI, 1990, Pump-and-treat method ineffective for contaminated groundwater, *HMCRI Focus*, November, p. 3.
2 Delphi Corp., 701 Haines Avenue NW, Albuquerque, NM, 87102.

3. Raymond, R.L., Jamison, V.W., and Hudson, J.O., 1975, Biodegradation of high-octane gasoline in groundwater, *Dev. Ind. Microb.* 16, Washington, D.C.
4. Raymond, R.L., Hudson, J.O., and Jamison, V.W., 1977, Bacterial Growth in and Penetration of Consolidated and Unconsolidated Sands Containing Gasoline, API Publication No. 4426, American Petroleum Institute, Washington, D.C.
5. Raymond, R.L. et al., 1978, Field Application of Subsurface Biodegradation of Hydrocarbon in Sand Formations, API Project No. 307-77, Washington, D.C.,

CHAPTER 6

Anaerobic Biodegradation: Sans Oxygen

Although under most situations virtually the entire subsurface is aerobic, isolated areas can reach anaerobic conditions. Even so, there have been few attempts in the realm of commercial bioremediation to utilize anaerobic subsurface conditions for site reclamation. In fact, the mere mention of the word "anaerobic" puts panic in the minds of potential clients and regulators. It almost falls into the category of genetic engineering in terms of a less than favorable response. This is unfortunate. The single factor people seem to overlook is the broad spectrum usage of anaerobic techniques in the wastewater treatment industry for many years.

The anaerobic treatment segment of the bioremediation industry is growing and is finally receiving the type of attention it deserves. Anaerobic systems are effective and, when carefully controlled, can often outperform aerobic treatment systems in a variety of situations. Many of the key developments to come into the bioremediation marketplace over the next few years will involve anaerobic systems alone or in combination with aerobic and other systems. Anaerobic systems may prove to be critical in the development of practical treatment methodologies for such recalcitrant compounds as the chlorinated solvents, PCBs, and pesticides.[1]

As part of the general scope of this chapter, we will define the term anaerobic and look at different types of treatment systems. We will also attempt to compare them with aerobic treatment systems. Examples of anaerobic applications will be given in lieu of case history information, which at present is lacking across the industry.

PROCESS DEFINITION

Anaerobic degradation pathways are utilized by organisms (usually bacteria) that grow in the absence of oxygen and employ some other electron acceptor in their respiration process. These other electron acceptors are commonly nitrate, sulfate, carbon dioxide, and other organic compounds.[2] Some anaerobes are known for their one major unappealing characteristic — their ability to cause disease (known as pathogenicity). Anaerobes are also associated with hydrogen sulfide production (rotten egg odor), sewage smells, and the production of methane. In fact, many people think of anaerobes as the "landfill bacteria".

Anaerobes have been identified that have the ability to degrade a broad array of contaminants. Gasoline and phenols are two materials to which naturally occurring anaerobic bacteria have been exposed for years.[3] They have finally adapted and acclimated to the point where these products can be used as food sources. The flammability and toxicity of some of the by-products of anaerobic degradation (hydrogen, hydrogen sulfide, and methane) may be cause for concern, but in many cases water and CO_2 are produced instead. Even if methane is produced during a large-scale remediation, it is possible that this gas can be captured and realistically utilized as a product of commerce. The prospect of a by-product of a remediation helping to pay for the cleanup is not far fetched and certainly is an attractive proposition.

As we continue to look at what anaerobes are, it may be helpful to classify them in three distinct categories: strict, facultative, and microaerophilic.[4] Strict anaerobes can tolerate no oxygen and can be killed rapidly via exposure to oxygen. Facultative anaerobes prefer anaerobic conditions, but if oxygen is present, they can make use of it on a temporary basis. Microaerophiles require oxygen, but only in very modest concentrations. Today's anaerobic treatment methodologies make use of the strict anaerobes and the facultative anaerobes. These types of bacteria are fastidious or "finicky" enough, without having to worry about the stringent requirements of microaerophiles. From a practical standpoint, any system or component of a system that requires too much operator attention simply doesn't cut it in today's "hands off" approach to bioremediation.

ANAEROBIC DEGRADATION

In addition to understanding what anaerobic bacteria are, the basic concepts of how they metabolize various compounds are important. Figure 6-1 indicates the types of anaerobic degradation mechanisms. Anaerobic bacteria can utilize a broad range of compounds for energy production. These compounds include simple carbohydrates, amino acids, fatty acids, and lipids, complex pesticides, and aromatic constituents of fuel products. It would be incorrect to assume that anaerobic bacteria have a finite appetite, but at the same time, there's plenty we don't know about the diets of specific anaerobes.

There are a few key anaerobic processes which warrant a closer look in this chapter, as they relate directly to the breakdown mechanisms of different compounds. These are anaerobic respiration, fermentation, and methanogenesis.[5]

Anaerobic respiration is the means by which bacteria utilize nitrate or sulfate as the ultimate electron acceptor. In order to do this, bacteria must possess a special enzyme that catalyzes the reduction of nitrate and the oxidation of cytochrome enzymes that together facilitate energy production. The process of denitrification (where bacteria reduce nitrate to nitrite and then to nitric oxide, nitrous oxide, or nitrogen gas) is perhaps best known because of its use in the wastewater treatment industry.[6]

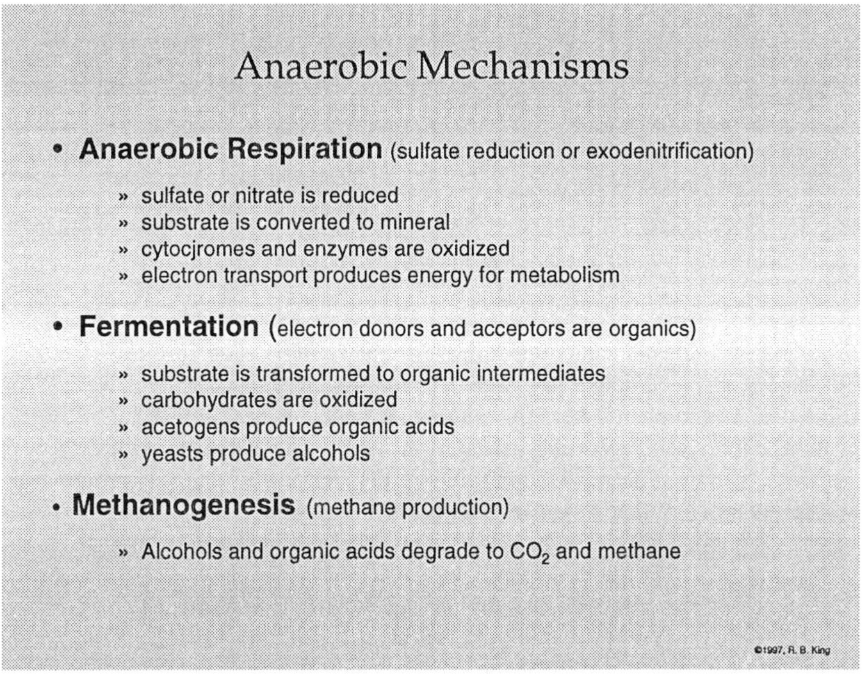

Anaerobic Mechanisms

- **Anaerobic Respiration** (sulfate reduction or exodenitrification)
  - » sulfate or nitrate is reduced
  - » substrate is converted to mineral
  - » cytocjromes and enzymes are oxidized
  - » electron transport produces energy for metabolism
- **Fermentation** (electron donors and acceptors are organics)
  - » substrate is transformed to organic intermediates
  - » carbohydrates are oxidized
  - » acetogens produce organic acids
  - » yeasts produce alcohols
- **Methanogenesis** (methane production)
  - » Alcohols and organic acids degrade to CO_2 and methane

Figure 6-1 Anaerobic mechanisms

The process of fermentation does not involve a respiratory chain or the use of cytochromes. Strict anaerobes and facultative anaerobic bacteria can carry out this process. Fermentation involves the breakdown of carbohydrates to glucose (a simple sugar) and various acids (pyruvic, lactic, acetic), and then to final products such as other acids, CO_2, alcohols, and gases. Many of these final products can be reclaimed, treated further in aerobic systems, or captured as an off gas. Although bacterial fermentation reactions and specific metabolic pathways are numerous, there is one key concept regarding the trade-off of energy consumed vs. energy created that needs to be understood. It happens that these bacteria typically get back double the energy that they put into the series of fermentation reactions. Thus, such a process is a very efficient one.

In the process of methanogenesis, specific types of bacteria act upon substrates such as alcohols, acetate, and formate in the production of CO_2 and methane.[5] This methane production is of great interest since it has the potential to be recovered and utilized as a fuel product. Methanogens have rather stringent growth requirements and are not necessarily reliable in terms of their ability to carry biochemical reactions through to completion.[7] Happily, they can do so in some instances and can carry a remediation to closure.

Before leaving this section of the chapter, it is important to note two key terms that play a role in the anaerobic side of treatment: cometabolism and dechlorination. Undoubtedly, we will see these terms (and their specific applications) more and more in the future. Cometabolism relates to a bacterium or bacterial consortium's

Table 6-1 Anaerobic Treatment Pros and Cons

Pros	Cons
High biomass production generates saleable by-products (methane, fertilizer)	Odor production
Excellent biomass retention	Potential pathogenicity of biomass
Minimal nutrient consumption	Significant acclimation period
Unique sludge characteristics	Temperature sensitive
Good buffering capacity	Limited number of anaerobic pathways
Good resistance to high organic loadings, metals, sodium	Ammonia sensitive

ability to break down a contaminant in the presence of an added primary substrate. This has become a particular key issue in the area of chlorinated solvent degradation where toluene and phenol have proven to be excellent cometabolites for TCE degradation, but there is recent discussion of anaerobic cometabolism with respect to pesticides. The term dechlorination refers to the cleavage of chlorine groups from a chemical structure that serves to detoxify that compound.[7]

PROS AND CONS OF ANAEROBIC TREATMENT

At the introduction to this chapter, basic statements were made regarding the characteristics of anaerobic systems. We will now look at the advantages and disadvantages of these systems in more detail. A summary of this section can be found in Table 6-1.

It has already been established that anaerobic treatment systems for use in remediation projects have not been well received.[1] On the positive side is the fact that anaerobic systems have been around for quite a while on the wastewater treatment marketplace. There are certain advantages associated with these systems. Development time and a track record are all that are needed to get anaerobic systems ready for field remediation work. The advantages of the anaerobic approach are quite strong and the disadvantages are specific issues that most probably impact cost more than any difficulty with the operation of a system.

The advantages of anaerobic treatment are that these systems do produce a higher volume of biomass than aerobic treatment systems. The integrity of the biomass tends to be so strong that infiltration by other types of microbes does not take place. This means that biomass produced is extremely efficient in its capacity to degrade a specific type of contaminant. The biomass produced in these systems typically has a high value as a commercial fertilizer due to its absorption of plant nutrients. Also, the biomass tends to stay in the system for long periods of time, allowing the system to efficiently handle high concentrations of waste. This can be accomplished without high consumption of nutrients. In essence, this lowers the operating costs of a system.

Other significant advantages of anaerobic systems include their high resistance to toxic constituents of the waste stream like metals and sodium, their key role as a pretreatment process, and the unique characteristics of anaerobic sludge which allow for design flexibility. In addition, these systems have excellent buffering capacity. There is no doubt that based on all these qualities, anaerobic treatment systems have a bright future.

Anaerobic systems have a down side as well. The disadvantages of anaerobic treatment systems include greater size and volume, which equates to a higher capital cost. Anaerobic systems also require a substantial acclimation period before they can efficiently treat the waste stream. This can be as long as 3 months. Anaerobic systems also tend to be temperature sensitive, responding in a qualitative fashion. It is important to point out that these treatment systems run in the temperature range of mesophilic organisms (20 to 45°C).

Other disadvantages to these systems include the generally stringent nutritional and environmental requirements of the bacteria making up these anaerobic consortia. Also, anaerobic systems may have difficulty in degrading complex waste streams due to the lack of anaerobic breakdown pathways for one or more compounds in the stream. These systems also cannot deal well with free ammonia in a system which can often be generated when there are proteinaceous materials in the waste stream. From an operational standpoint, the exchange for the system can be a costly venture.

TYPES OF ANAEROBIC SYSTEMS

Unlike their aerobic counterparts which have been discussed in other sections of this book, the anaerobic systems are few in type. In fact, much of the anaerobic system engineering technology is quite comparable. Through the rest of this section, we will look at some of the different types of anaerobic bioreactors and discuss both *ex situ* and *in situ* treatment systems.

There are a limited number of anaerobic bioreactors currently available. These include digesters, activated sludge systems, fixed film, fluidized bed, sludge blanket, and anaerobic rotating biological contactors (ARBCs). The fluidized bed and sludge blanket types of bioreactors are receiving a great deal of attention for use at remediation sites and these systems will be discussed in more detail than the other types of bioreactors.

Like all other biological treatment systems, contact between the waste, the biomass, and nutrients is essential. The mixing process, subtle as it may be in an anaerobic process, is an important engineering aspect to each system. Mixing ensures that the proper degree of contact takes place and that "dead zones" are minimized.

Table 6-2 illustrates the pros and cons of the different types of treatment approaches. A short description of some of the lesser used reactors follows with a lengthier discussion of the more popular reactor types.

Digesters

These are the simplest of all the anaerobic treatment systems. Digesters utilize a mixed tank system which is run in a continuous or batch mode. Anaerobic digesters

Table 6-2 Pros and Cons of Specific Anaerobic Approaches

Treatment approach	Pros	Cons
Digester	Simple operation	Requires heating
	Proven track record	Long retention times
Activated sludge	Good BOD/COD reduction	Volatile emissions
	Excellent pretreatment	Temperature sensitive
Fixed film	Excellent tertiary treatment	Prone to biofouling
	Good filter for gas, solids, dissolved organics	Difficulty with high organic loading
Rotating biological contactor	Shock resistant	Large size requirement
	Supports lush biomass	High capital cost
Fluidized bed	Good performance at high flow and high organic loading	Variable retention times
	Resists biofouling	High capital cost
Sludge blanket	Utilizes unique sludge characteristics	Prone to form "dead spots"
	High removal rate	
	Short residence time	Plugging by gases
Soil cells	Treats recalcitrant compounds	Difficult to maintain anaerobic conditions
	Compact design	Unproven record
In situ	Nitrate is efficient electron acceptor	Permitting sometimes difficult
	Attractive costs	Stringent pH/redox control required

are typically heated. They are most often found in municipal service and are subject to extensive hydraulic retention times.

Anaerobic Activated Sludge

A slightly more sophisticated technology is the activated sludge system. Typically, this system employs two tanks where the first is used for treatment much in the same manner as a digester. A second tank is used to thicken the sludge before it is returned to the front end of treatment train. These systems allow good control of the biomass density, but are subject to control problems with volatiles and difficulties in temperature control. There have been problems with the sludge thickening process, as overproduction of gases in the second tank can cause the sludge to flocculate or bulk to the top of the tank.[5] This sludge bulking hampers the recycling of the sludge back to the treatment tank.

Fixed-Film Units

One of the more significant breakthroughs in anaerobic treatment was the development of a fixed-film reactor. Aerobic fixed-film reactors are extremely efficient, often problem free, and are quite popular at this time. In fact, trickling filters (a related approach) have been very successful in wastewater treatment in both the municipal and industrial markets.[6] The anaerobic counterpart has not as quickly caught on, but the establishment of the technology has stimulated the development of other practical anaerobic treatment processes.

Typically, these reactors contain rock media or plastic saddles which provide points of attachment for the bacteria. The waste stream enters the reactor from the base and passes upward through the media. Contaminants are trapped in the biomass and on the media and become available for treatment. The media also serves as a biofilter for gases and suspended solids. This system retains its biomass for long periods of time, but does have a propensity to plugging.

Anaerobic Rotating Biological Contactors (ARBCs)

The early 1980s saw the development of an enclosed unit with a series of anaerobic compartments.[5] This system contains a number of inert rotating attachment disks lined up side by side in each compartment and turned by a common shaft. As the waste enters the unit, it comes into contact with the microorganisms attached to the disks which remain partially submerged at all times. Through the mixing and turning action, the wastes are degraded as a result of being in intimate contact with the biomass. These systems have gained some acceptance on the municipal level, but are limited in application due to their large size and overall capital cost.

Fluidized Bed Bioreactors

One of the newest state-of-the-art anaerobic treatment reactors is the fluidized bed or expanded bed reactor.[5] The basic design concept entails the passing of waste from the base of the reactor up through a bed of particulate media, causing the fluid movement and expansion of that bed. Thus, a maximum surface area for microbial attachment and particle adsorption is created. These units have been demonstrated at hundreds of locations nationwide and have operated in aerobic, anaerobic, and facultative anaerobic modes.

Fluidized bed reactors are known for their ability to deal with high BOD/COD loadings. One of the systems produced by Envirex of Waukesha, WI offers the ability to treat wastes of high flow rate/high organic loadings. These units have large upright volume in order to maximize space utilization. The standard flow path in this system mixes the waste stream with recycled treated water entering the front end of the system. Treated water and effluent gas are removed from the system and are treated further, recycled, or discharged. An external pump system aids in the bed fluidization and recycle of sheared media and biomass. Most often, these systems have no internal moving parts to worry about and do not plug due to solids buildup or gas mobilization (bulking) of the sludge.

Fluidized bed reactors resist abrupt changes in flow and waste concentration. They offer reasonable retention times at a reasonable price, as long as the scenario is one of high flow rate/high organic loading. These units are not known to be efficient for low flow/low organic loading situations. Like most bioreactors today, these units are designed to operate relatively maintenance free and require only minimal nutrient addition.

Upflow Anaerobic Sludge Blanket (UASB)

Perhaps the most interesting of all the available anaerobic treatment systems is the upflow anaerobic sludge blanket.[8] This is a unique type of system that is compartmentalized and is capable of handling a wide variety of specific sludge characteristics presented by a given waste stream. Usually, there are three compartments within this type of reactor. These are the sludge bed, the sludge blanket, and the separation zone.[10]

The sludge bed lies in the base of the reactor. The waste is passed though this bed under minimal agitation. Uniform distribution of sludge and waste ensures intimate contact and maximum treatment. This compartment makes up about one third of the reactor volume and provides the bulk of the treatment because it is the point of maximum contact between the waste, the bacteria, and nutrients. Rising gas bubbles produced during treatment in this compartment serve as a "natural" mixing mechanism. One drawback is that this zone can be prone to "dead spots" and channeling which can obviously reduce treatment efficiency.

The second compartment is the sludge blanket. This compartment occupies about 60% of the reactor volume. It contains highly flocculated sludge and minute gas bubbles which aid in achieving ideal mixing and provide the bulk of the remaining treatment.

The separation zone takes up the balance of the volume within the reactor. This area typically has a capture apparatus series which collects the gas and releases any biomass that may be attached to gas bubbles. Also, other solids are removed and treated effluent is allowed to exit the reactor.

These units are known for high removal rates of 90% or greater, short residence times on the order of 3 hr, and conversion reactions that not only reduce organic loading, but generate a recoverable gas product. Like the fluidized bed reactors, these units have no internal moving parts and are energy efficient. There are some significant advantages to UASB units. They produce far less sludge than aerobic units of comparable size, they acclimate quicker than other anaerobic units (5 to 6 weeks), they are not sensitive to changes in flow rate and strength, and they can withstand months of down time with little detriment to the system.

EX SITU SOIL TREATMENT (SOIL CELLS)

The treatment of contaminated soils in piles or pits is a well-established technology under aerobic conditions. Most recently there has been great interest in evaluating anaerobic treatment via this route, especially for contaminants that can undergo efficient anaerobic degradation (such as pesticides). The general concept involves the enclosure of the pile or covering of the pit in what amounts to an airtight bag. This is quite similar to the anaerobic glove bags utilized in laboratories. In addition, the soil is saturated through the introduction of water, and special oxygen-scavenging cultures of bacteria are added to optimize the anaerobic conditions. Once this is accomplished, the naturally occurring anaerobic bacteria can then proliferate and ultimately act upon the contaminants. One disadvantage is that temperature and moisture excursions can cause the resident actinomycete population to generate antibiotics which can create a toxic environment.

One can envision this type of an approach being applicable for soils excavated from a citrus grove or from some other agricultural application. Also, any site contaminated with chlorinated solvents might lend itself, at least in part, to this type of a treatment approach, with the possibility of methane recovery as a by-product from the system. Ultimately, the biodegradation of the higher PCB congeners might pass through processes such as these as a pretreatment to aerobic biodegradation.[9]

IN SITU ANAEROBIC REMEDIATION

Classic *in situ* treatment has been conducted as an aerobic process even though, as the system is often implemented, the conditions at the site may reach an anaerobic state or have just a trace of oxygen present. Many questions still remain regarding the broad spectrum applicability of an *in situ* anaerobic treatment system, but in theory such systems should be possible. Lee et al.[12] discuss the use of alternate electron acceptors such as nitrate in combination with pH and redox potential adjustments in order to control anaerobic activity *in situ*. They clearly point out that the processes of denitrification, sulfate reduction, and methanogenesis are ongoing in the subsurface at low kinetic rates and can be stimulated by the addition of low levels of nutrients. These anaerobic processes are low energy producers and therefore produce limited biomass, which quite interestingly limits biofouling of the system.

An *in situ* anaerobic treatment process will be slow and may be questionable in terms of its overall efficiency in meeting regulatory treatment levels. Additional work continues with an emphasis on sequenced aerobic and anaerobic systems.[9] Success in treating a variety of recalcitrant compounds in the laboratory via anaerobic pathways will ultimately dictate the true commercial emergence of this specific treatment approach. Figure 6-2 illustrates the evolution and current status of anaerobic treatment and points the way for future anoxic systems development.

DESIGN CONSIDERATIONS

The various design considerations for anaerobic systems are no different than for aerobic systems regardless of whether one is evaluating a bioreactor, an *ex situ* soils treatment, or an *in situ* system. The overriding principle is that one must ensure that optimum contact occurs between microbes, contaminants, and nutrients. The appropriate cometabolite or electron acceptor must be present and available in order to facilitate degradative activity.

Testing is a significant requirement in any application of anaerobic treatment. A general characterization of the anaerobic bacterial population is important from the standpoint of identifying specific biodegradation functions and their associated requirements, and understanding potential pathogenicity issues associated with the microbes. The pH, temperature, and redox potential values are important in terms of manipulating the optimal growth environment for the bacteria. Total suspended solids, nutrient concentrations, and COD/BOD loadings are important in terms of sizing primary equipment and accessory items that must be present to fine tune the system.

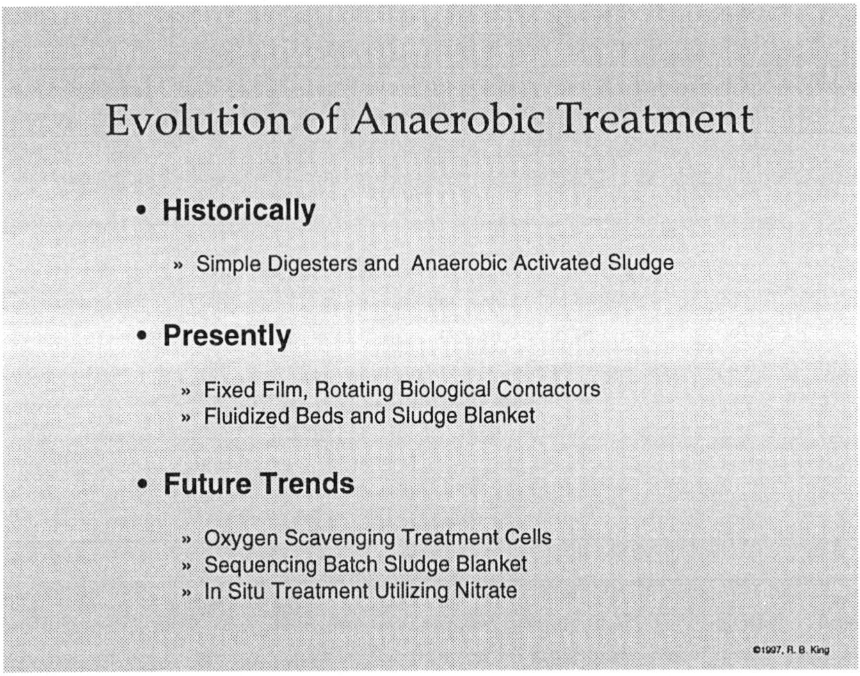

Evolution of Anaerobic Treatment

- **Historically**
  - » Simple Digesters and Anaerobic Activated Sludge
- **Presently**
  - » Fixed Film, Rotating Biological Contactors
  - » Fluidized Beds and Sludge Blanket
- **Future Trends**
  - » Oxygen Scavenging Treatment Cells
  - » Sequencing Batch Sludge Blanket
  - » In Situ Treatment Utilizing Nitrate

Figure 6-2 Evolution of anaerobic treatment

In addition to laboratory-generated information, the layout of the site, the specific space requirements, and the level of activity of the facility become important issues in designing a system. Performance history often sets the confidence level of the degree of treatment the system might attain. Three significant issues to be considered during the design phase of a project are expected project duration, negotiated or regulated bioremedial end point or closure concentrations, the system's ability to achieve these, and costs.

APPLICATIONS FOR ANAEROBIC TREATMENT

Despite a limited number of case histories on anaerobic treatment remediations, there are some innovative applications that are just now being addressed. The list of potential future applications stimulates some interesting predictions as to where the technology might be useful:

- The treatment of pesticide-contaminated soils where an initial dechlorination step is necessary (This process may be combined with oxygen scavenging activities and perhaps an aerobic polishing step.)
- The treatment of BTEX-contaminated aquifers where nitrate is utilized as the electron acceptor
- The treatment of municipal sludge in order to generate methane which can be used as a source of electrical power generation

- The treatment of process waste from chemical and pharmaceutical manufacture characterized by high organic loadings and made up of mixed organic solvents
- The treatment of complex process wastes from tanneries, pulp and paper mills, and assorted other manufacturing operations
- Waste streams that contain high nitrate concentrations
- The treatment of high organic loadings from canning, bottling, baking, and other food processing operations
- Concentrated alcohol waste streams from brewing and distilling
- Treatment of wastes from new fuel product operations
- Treatment of landfill leachate with an emphasis on methane recovery

This amounts to an impressive list of potential opportunities for anaerobic treatment systems. Their ultimate implementation will rely heavily upon the progress of research and development activities in the area of treatment technology, and on the initiative of practitioners in the field who are not limited to implementation of only aerobic treatment solutions. The next 2 to 3 years will likely see the beginning of a whole new era of biological treatment technology highlighted by development of well-founded anaerobic microbiological methods and high quality, reliable process applications with exciting new potential.

REFERENCES

1. Calmbacher, C.W., 1991, Biological treatment gaining acceptance, *Environ. Protection,* 2, 2, 38–41.
2. Pelczar, J.R., Reid, R.D., and Chan, E.C.S., 1977, *Microbiology,* Fourth Edition, Chapter 4, McGraw-Hill, New York.
3. Ehrlich, G.G., Goerlitz, D.F., Godsy, E.M., and Hult, M.F., 1982, Degradation of phenolic contaminants in ground water by anaerobic bacteria: St. Louis, Park, MN, *Ground Water,* 20, 6, 703–710.
4. Moat, A.G., 1979, *Microbial Physiology,* Chapter 3, John Wiley & Sons, New York.
5. Obayashi, A.W. and Gorgan, J.M., 1985, *Management of Industrial Pollutants by Anaerobic Processes*, Lewis Publishers, Chelsea, MI, pp. 1–39.
6. Grady, C.P.L., Jr., and Lin, H.C., 1980, *Biological Wastewater Treatment*, pp. 833–876.
7. Chakrabarty, A.M., 1982, *Biodegradation and Detoxification of Environmental Pollutants*, CRC Press, Boca Raton, FL, pp. 127–140.
8. Biothane Corporation, 1988, *Biothane Digest,* Camden, NJ.
9. U.S. Environmental Protection Agency, USEPA Report 600/9-90/041, Office of Research and Development, Washington, D.C. 20460.
10. Kennedy, K.J., Sanchez, W.A., Hamoda, M.F., and Droste, R.L., 1991, Performance of anaerobic sludge blanket sequencing batch reactors, *Res. J. Water Pollution Control Fed.,* 63, 75–83.
11. Lanzarone, N.A. and McCarty, P.L., 1990, Column studies on methanotrophic degradation of trichloroethene and 1,2-dichloroethane, *Ground Water,* 28(6), 910–919.
12. Lee, M.D., Thomas, J.M., Borden, R.C., Bedient, P.B., Ward, C.H., and Wilson, J.T., 1988, Biorestoration of aquifers contaminated with organic compounds, *Crit. Rev. Environ. Control,* 18(1), 52–55.

CHAPTER 7

Surface Bioremediation of Soils and Sludges: Land Treatment

Land Treatment is a generic term coined by RCRA to cover the controlled application of bioremediation to surface soil and sludge piles. The technology is a direct outgrowth of the petroleum refinery practice of land farming. While land farming as originally practiced is no longer permitted without appropriate containment, its history provides the earliest evidence of the use of aerobic microbial processes to provide remediation of petroleum products.

In the early 1900s, when spewing smokestacks were considered a sign of progress, petroleum refineries were not very careful about their waste disposal practices. The usual removal method for several of the sludges generated in refining processes was to dump them into sandy soil, where they seemed to go away of their own accord (or at least the volume of sludge was reduced). This dumping was often done on the banks of the local river or streams supplying process water for the refinery. Many former and current refinery sites still show the soil-staining characteristic of this activity.

The science behind all this was virtually unknown at the time; indeed, some fraction of the material leached into the water and was removed from the "treatment" area before any microbial activity could do much to change the chemical structure of the sludge components. Another fraction volatilized into the air and was carried off by the local wind in a relatively unchanged chemical state. Since at least some of the sludge was acidic, it is reasonable to assume that the removal of any metals occurred by leaching. This treatment was used for API separator sludges, heat exchanger bundle sludges, tank bottoms, and other phase-separated sludges that eventually were listed as *Category K* wastes under RCRA.

Gradually over the decades, some understanding of the reasons for the reduction in sludge volume was gained, and attempts at optimization began. Fertilizers were added to promote the sludge volume reduction, water management and pH control were begun, and the use of spray equipment and tillers specifically designed to apply the sludge and aerate the soil became the industry standard. Samples were monitored for pH, bacterial nutrients, and sometimes bacterial populations and the test results were used to determine when more sludge could be added. Improvements to the process made the sludge disappearance go faster and more completely; the technology was becoming a common and reliable means of sludge disposal (or at least provided significant volume reduction).

In the 1970s, this practice began to be discouraged as the first wave of stiffer environmental regulations came into being. The regulations growing out of RCRA provided a tough regulatory framework and specified standards for the construction and operation of *land treatment* systems. The new regulations redefined the former land-farming practices and specified improved methods of control.

Most of the improved control methods dealt with the issues of environmental fate of the contaminants and any possibility for their transport away from the treatment area. The former practices had encouraged the leaching of wastes into surface or groundwater and volatilization as methods to reduce the contaminant loading, whereas current practice curtails or eliminates both of these removal mechanisms.

As practiced by the hazardous waste treatment industry, land treatment proceeds according to a different intent from historical land farming; whereas formerly, sludge was applied to soil which initially was clean, the waste treatment approach objective is to clean dirty soil. This distinction, while seemingly minor, explains some of the peculiar permit requirements for the process.

REGULATIONS

The specific part of the Code of Federal Regulations (CFR) dealing with land treatment is found in Title 40, parts 260 and 264. In Part 260.1, a land treatment facility is defined as "*a facility or part of a facility at which hazardous waste is applied onto or incorporated into the soil surface; such facilities are disposal facilities if the waste will remain after closure.*"

Part 264, Subpart M specifically deals with the regulation of land treatment operations. This subpart describes the construction, testing, and operating standards required to treat RCRA hazardous waste materials applied to soil in a treatment cell. If the waste material is to remain in the treatment cell after completion of the cleanup and is not removed (closed in place), more stringent closure standards are applied to insure that the material is completely transformed or encapsulated.

As indicated above, this treatment method is primarily applicable to organic sludges and semisolid organic materials. Volatile materials would not be treated using this approach because the majority of material would volatilize into the atmosphere before it had a chance to degrade.

Given the attention to detail required by the regulations, the first item to be determined is whether the waste solids are considered hazardous under RCRA. If a specific RCRA classification applies to the waste, then the construction and monitoring details specified in 40 CFR Part 264 must be followed. These specifications include impermeable double lining systems with monitoring devices in the space between the liners; these items can add substantially to the total cost of the remediation.

However, many hydrocarbon or fuel wastes which require remediation do not meet the definition of hazardous under RCRA. In the absence of specific chemical compounds on the hazardous substances lists which were discharged as purified chemicals, many sites where fuel or blended refinery products were spilled or released do not require the stringencies of the RCRA standards.

As a practical matter, most state agencies will require that containment of the solids is designed and installed so that possible migration of liquid contaminants into the soil or groundwater below the contaminated material is minimized. Thus some form of impermeable liner system is almost always required, and a prepared pad or cell is constructed for containment of the soil to be treated. The cost of the construction of this cell is usually a substantial fraction of the total project cost. This cost is more bearable when it can be spread over several "batches" of soil or sludge contaminated with organic material; the treated material may sometimes be used as fill on the site where the material was generated, and treatment on the generation site may simplify the permitting requirements.

In addition to meeting construction standards, the treatment cell will often have to be operated under some form of operating permit(s), usually granted by the state. These requirements vary considerably from state to state, and you will have to ask the agency for guidance on which permits must be applied for. Typical permits include air or emission limit permits (often including air monitoring as part of the operations), discharge permits (even though there is normally no discharge from the cell, some states require this form of permit and also require groundwater monitoring downgradient of the treatment cell), and an operating permit which usually gives the agency permission to watch the process and tell you when you may stop. Some permits require that the municipality sign part of the application, and this process may require months to complete in some cases (see Figure 7-1).

FEASIBILITY TESTING

Given the apparent simplicity of the technique, it is not surprising that the types of tests to determine whether land treatment will work are fairly simple. Typically, soil samples are tested for ammonium nitrogen and o-phosphate content, bacterial enumerations of total heterotrophic bacteria and specific hydrocarbon degraders, soil pH and buffer capacity, and water content or field capacity while the hydrocarbon content is being determined. It is recommended that a rough idea of the variability of these parameters be obtained while the delineation of the extent of remediation required is performed, although some mixing is permitted while the soil or sludge is being staged for placement in the treatment cell. It should also be stressed that this technique, while conceptually straightforward, has no guarantee of success after a couple of tests! The operations still require monitoring, process fine-tuning, and vigilance to get the solids to the cleanup standard.

Typically, a brief microcosm stimulation test is run by adding nutrients to the contaminated soil to bring the approximate nutrient ratio of carbon to nitrogen to phosphorus (C:N:P) to 400:10:1. This ratio was empirically used by land farmers for many years. The soil pH is adjusted if necessary to a range of 6.5 to 8.0, typically

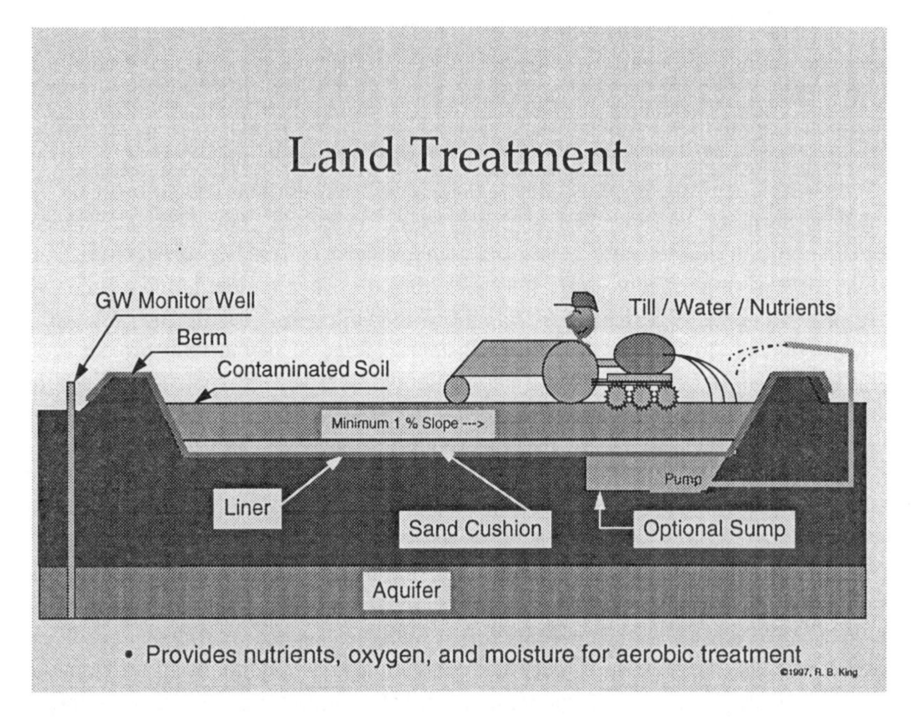

Figure 7-1 Land treatment

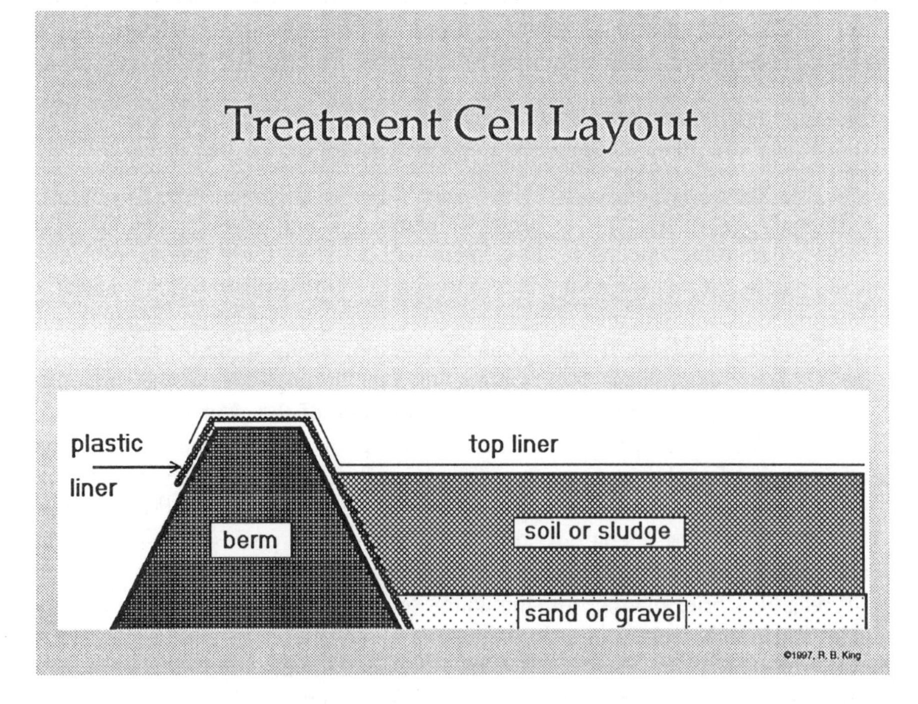

Figure 7-2 Treatment cell layout

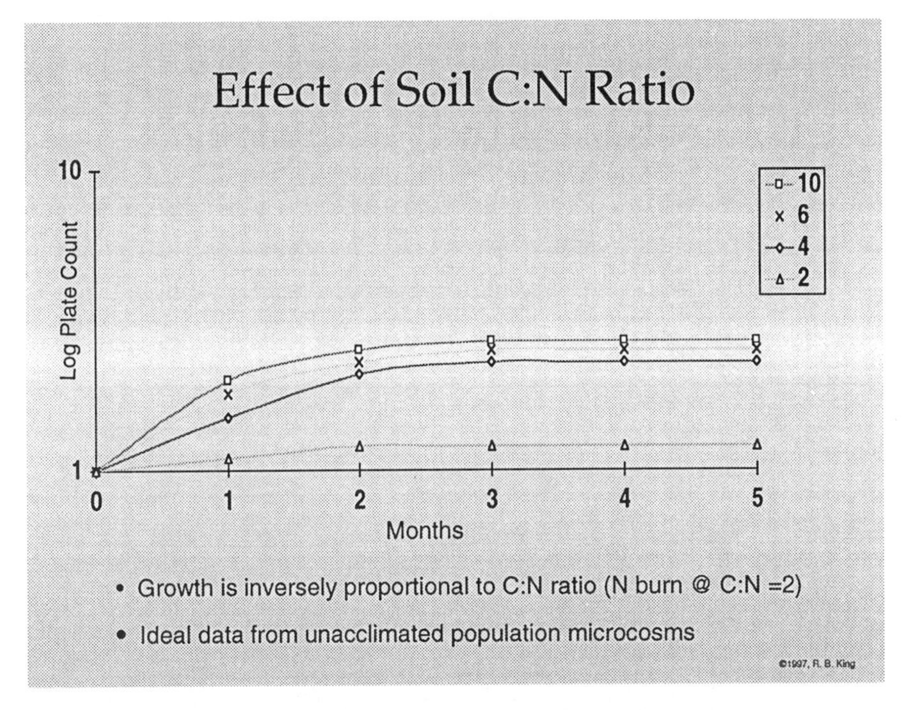

Figure 7-3 Effect of soil C:N ratio

using crushed limestone. If necessary, water is added to bring the soil to approximately 50 to 80% of its field capacity (saturated water content of unsupported soil). After a period of 1 to 4 weeks, the soil is tested for hydrocarbon content (usually indicated by TPH or some specific compound) to determine the effect of favorable conditions on the ability of the native organisms to degrade the contaminants. Unless some other toxic compounds are present, this test usually indicates that bioremediation will degrade the material present at least to some degree. The optimal nutrient requirement for the microbial consortium during field treatment will become apparent with a simple comparison between microcosm treatments. Varying the nutrients between microcosms will indicate the best ratio for full-scale operation.

In theoretical terms, the C:N ratio should be optimal when it approximates the 400:10 "textbook" number. Lab testing of unacclimated populations in microcosms will bear this out (see Figure 7-3). Decreasing the C:N ratio (by increasing NH_3 concentration) is observed to slow the rate of growth and results in lower ultimate population numbers. At moderate ratios, nitrogen addition is beneficial. However, when the ratio reaches around 2 (N = 50% of C) the microcosm cannot even get started due to nitrogen burn. The same thing happens when yard grass turns yellow after being overfertilized.

On the other hand, a well-acclimated microbial population might behave differently, as conditions in the field are often somewhat different. Figure 7-4 depicts the data from a land treatment project containing #6 fuel oil in soil that had been in place 100 years. These bacteria were found to be able to handle a C:N of 2.5 (N at

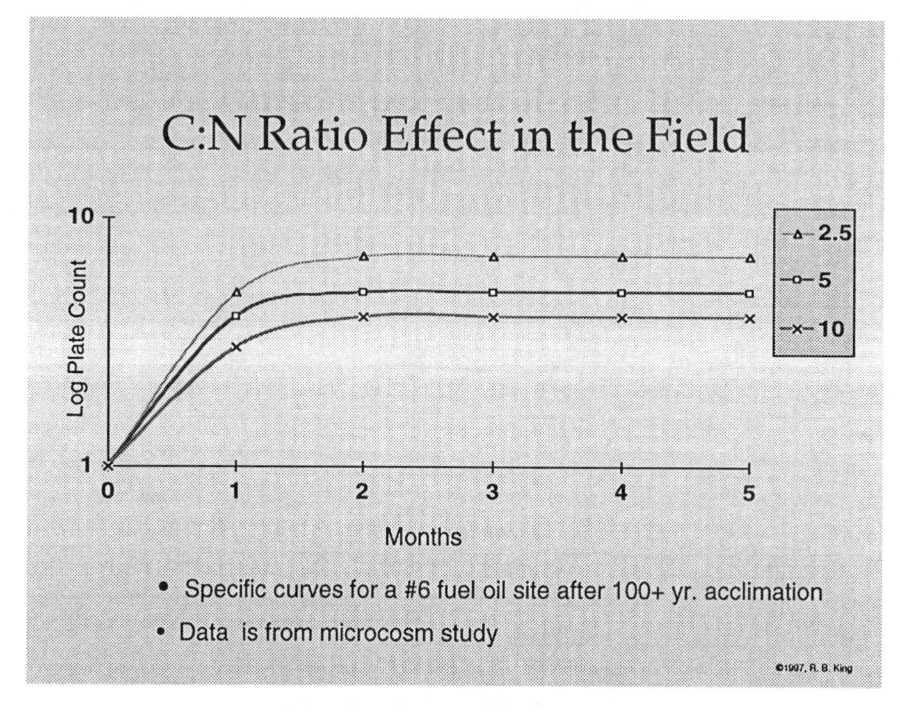

Figure 7-4 C:N ratio effect in the field

40% of C). The degradation rate was highest at a C:N near the burn point for laboratory microbes. This instance shows the importance of proper testing prior to final project design. The application of a textbook C:N ratio would have doubled the time to closure. So, the C:N ratio (or application rate) needs to be determined during treatability testing for correctly estimating timing and costs.

These tests may become more elaborate if the desire is to detail the degradation pathway (which would require detailed GC/MS or other analyses frequently) or to determine the extent of remediation possible for heavily contaminated soil. For refractory compounds, land treatment offers an easy pathway to introduce compound-specific degrading organisms (designer bacteria) into the soil; the appropriate tests for these bacteria would also need to be run to determine their initial survival and the impact of the exogenous microbes on the degradation rate and degree.

The most important factor in the testing protocol is that the compound which will be used as the judgment criterion for success be tested and, if possible, shown to be treatable to the site cleanup standard.

DESIGN AND CONSTRUCTION

Once the land treatment process has been shown to be feasible, the full system must be designed. The design methods below DO NOT apply to RCRA sites (see 40 CFR 264 for those specifications). The methods given below have generally been approved by the appropriate regulators, the necessary operating permits have been granted, and the process has been managed to regulatory completion (i.e., permission to stop). It still is worth a call to the agency involved with the site to determine what the permit and design requirements are for a given site and state.

The first requirement is an accurate estimate of the volume of soil to be treated. This is usually determined as a part of the site investigation, when the samples for the treatability tests are taken. If the volume estimate is based on the volume of soil in the ground, a factor from 1.25 to 1.4 should be multiplied by the in-ground volume to account for "fluffing" which occurs during excavation and tilling. The soil texture should also be noted at this time, as the amount of berm construction required may be affected by the soil density and its natural tilth.

The basic principles of the design are

- Provide enough surface area to spread the total fluffed soil up to 18 in. deep on an impermeable surface or liner system.
- Provide a slight (0.5 to 1%) slope for drainage across the cell to a sump.
- Provide a collection sump at the low end of the cell to collect rain water runoff from the cell.
- Provide retaining berm walls to keep the soil or sludge contained and prevent water from outside the berms from entering the cell.
- Provide access to the cell via a ramp for tilling equipment, nutrient-addition equipment, and water-pumping equipment.
- Provide a sand or other readily drained base material over the liner to protect it from the tilling equipment and promote drainage of excess water to the sump.
- Provide an easily removed cover if needed to minimize the amount of storm water which requires management in the cell.

A relatively flat area should be selected to meet the first two of these principles. Although a relatively large amount of surface is needed, piling the spread soil to a depth of greater than 24 in. usually reduces the tilling, and hence aeration, efficiency considerably and will result in a longer remediation time. A typical treatment cell layout is shown in Figures 7-1 and 7-2.

If the soil is clay to a depth considerably greater than the depth of surface contamination, it may be possible to obtain permission to perform the treatment without constructing a liner system. The basis for this is the presence of an impermeable surface below the contamination, eliminating the need to construct one. This type of soil structure may be found in the Piedmont area of the Southeast and the Gulf Coast, among others. The agency may require that piezometers be installed below the treatment zone to insure that contaminants are not migrating toward the groundwater.

An alternative which may be used to eliminate or minimize the liner system requirements is the conduct of detailed fate and transport calculations on highly humic soils, where the rate of transport (percolation) through the humic material is slower than the degradation rate. This has been successfully demonstrated on at least one EPA Superfund site in Region I (Iron Horse Park), but considerable testing was required to demonstrate that these conditions applied. At this site, the contaminants were almost exclusively polyaromatic hydrocarbons (PAHs), multiple ring aromatic compounds with relatively low solubility in water, low volatility at ambient temperatures, and a high sorption coefficient for humic soils. The elimination of the liner saved the PRP approximately one third of the estimated construction cost for the land treatment system.

Assuming that the soil is not clay and will require excavation, the large, flat area selected for the cell will need an impermeable floor. The possible liner materials include concrete, asphalt, plastic liners, clay, or some combination of these. Paved parking lots in good condition make appropriate liner systems. Plastic liners usually must be 60 to 80 mil in thickness to absorb the shock of the tilling equipment. Compacted clay should be at least 4 in. in thickness to meet most state regulations.

The slope can usually be built into the construction grading plan while the equipment for excavation is on hand. An area at the low point of the grading needs to be reserved for the sump.

If a cover is not specified for the treatment area, the collection sump should be designed to hold all of the water from a 25-year rain event across the entire cell. Some of the field capacity of the drainage material (usually sand or gravel) may be included in the capacity calculation for the sump. The height of the collected water should not rise above 6 in. from the top of the retaining berm in sizing the sump area. This sump should be sufficient to keep water from leaving the cell through runoff. The other purpose for the sump is to collect sufficient water for irrigation during dry periods. A typical sump layout is shown in Figure 7-5.

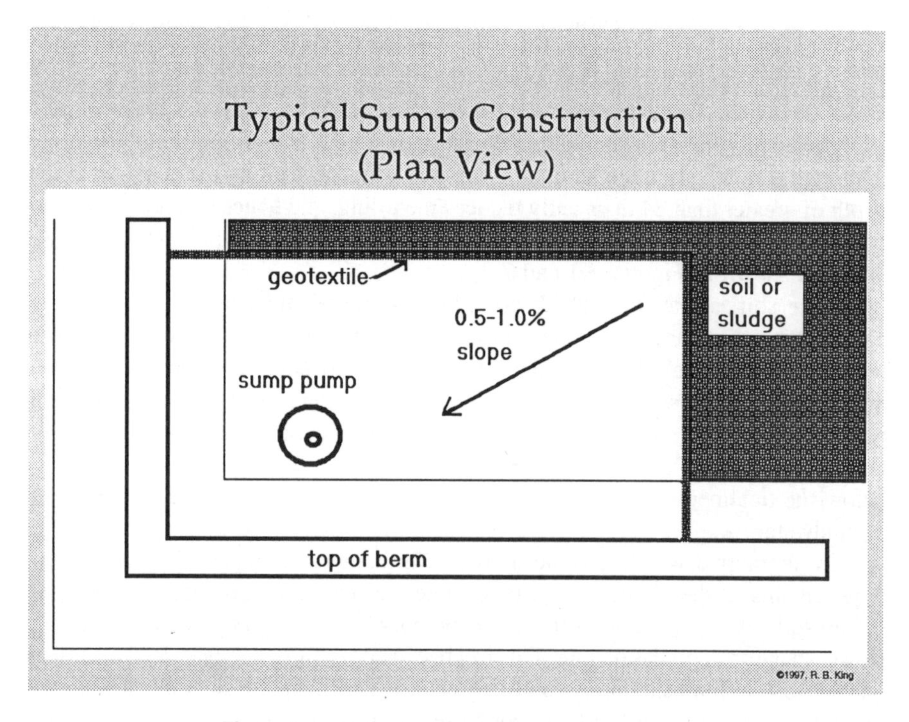

Figure 7-5 Typical sump construction (plan view)

The hydraulic loading on the sump may be greatly reduced by specifying a plastic cover for the treatment area. A lightweight composite plastic, typically 6 mil or so in thickness, is solvent-welded together to the appropriate size to drape across the entire cell, and is tied down at several points along the berms to hold it in place. When the cell requires tilling or irrigation, the liner is removed for these operations by rolling it up along one edge. At the completion of these activities, the liner is placed back over the cell and tied down. The tie-down should not prevent ambient air from entering into the cell, because the intent is to keep the process aerobic. The sump still must have enough capacity to irrigate the entire cell once, typically to an equivalent of 1 in. of water over the entire treatment area. If the sump area required for this amount of water is too large for the available area, the water may be stored in portable tanks outside the treatment cell and pumped back for irrigation when necessary. This approach does require additional expense for tank rental and pumping equipment. At one large land treatment cell, six 20,000-gal portable tanks were used for 3 months in this service.

In the authors' experience, uncovered land treatment cells which must operate for more than 4 months may have problems with the hydraulic loading on earthen berms. Heavy rain events have washed out earthen berm walls when the water piled up at the low corner to the top of the berm, especially in large cells (more than about one acre). In one case, approximately 500,000 gal of impounded water was released into a navigable waterway! Fortunately, the water had been tested the week before the release and had been demonstrated to be clean.

The water retainage may be terraced to prevent all the water from reaching the sump, but this may make irrigation pump placement more complicated. In essence, this approach creates a series of sumps to distribute the hydraulic load, and has been used only when one sump cannot handle the water loading (as described above). It is simpler to cover the cell with a liner after the sump has enough water to provide irrigation and manage the water in a closed recirculation system.

The berm walls may be constructed of clean fill, concrete, or other material. The design height is usually 12 in. above the planned top of the contaminated soil; since this soil is placed on a drainage base, the height of the base must be included in the calculation. For example, a cell with a 6-in.-thick sand base and 12 in. of contaminated soil would need a berm wall 30 in. high. If earth walls are used, they should be compacted and "buttered" with clay on the inside surfaces to prevent seepage of contaminants through them, if plastic liner materials are not being used inside the berms. The clay layer should be about 2 in. in thickness and must be tapered into the bottom liner to prevent leakage through the joint. On the upslope side, consideration may be given to coating the outside surface with clay to prevent seepage into the cell. The slope of the berm walls should not exceed 45°. It is useful to maintain a walkway on the top for access purposes. Figure 7-6 illustrates a typical earthen berm and liner construction detail.

The equipment access ramp is usually positioned on the side of the cell closest to the street or access gate (see Figure 7-2). It should not be close to the sump, so that equipment can still get onto the cell after a rain event.

The sump pumping equipment should be sized to provide water to the cell during dry seasons. This is empirically estimated based on the soil drainage properties, and is often based on providing 1 in. of water across the entire cell in 8 hours. The water

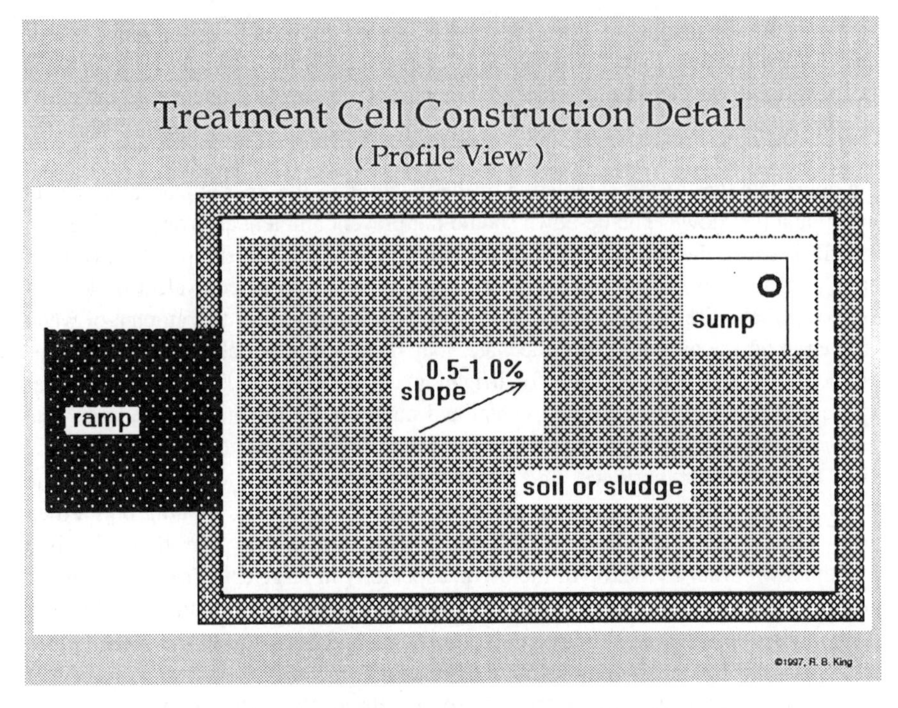

Figure 7-6 Treatment cell construction detail (profile view)

is usually sprinkled across the treatment area using garden-type sprinkling equipment (soaker hoses or lawn sprinklers). This equipment must be easily moved so that tilling can be conducted without destroying it.

The drainage material (sand base) serves two purposes: to drain water from the base of the contaminated soil to keep the soil from becoming saturated, and to protect the liner material from the effects of the tilling equipment. The water should be readily conducted from the base of the soil to the sump for collection and redistribution. The drainage material can be clean coarse sand or gravel, normally spread from 6 to 12 in. in depth across the entire pad except for the sump area. Relatively permeable geotextile material may be placed along the edge of the sump area to keep the sand or gravel out of the sump.

The tilling equipment should be selected to till the entire area in less than 1 day. The size may range from a garden rototiller to a tractor and 6-ft-wide disk set. The tradeoff between equipment cost and labor cost should guide the sizing calculations. The "standard" disk set provides reasonable loft to the soil but is not an efficient mixer; rototillers do a better job of incorporating nutrients, limestone, and other additives into the solid matrix and in drying excess water after rain events.

NUTRIENT APPLICATION

Once the construction of the treatment unit is complete, soils may be layed in. The lift height is optimal somewhere between 12 and 24 in. of contaminated materials for

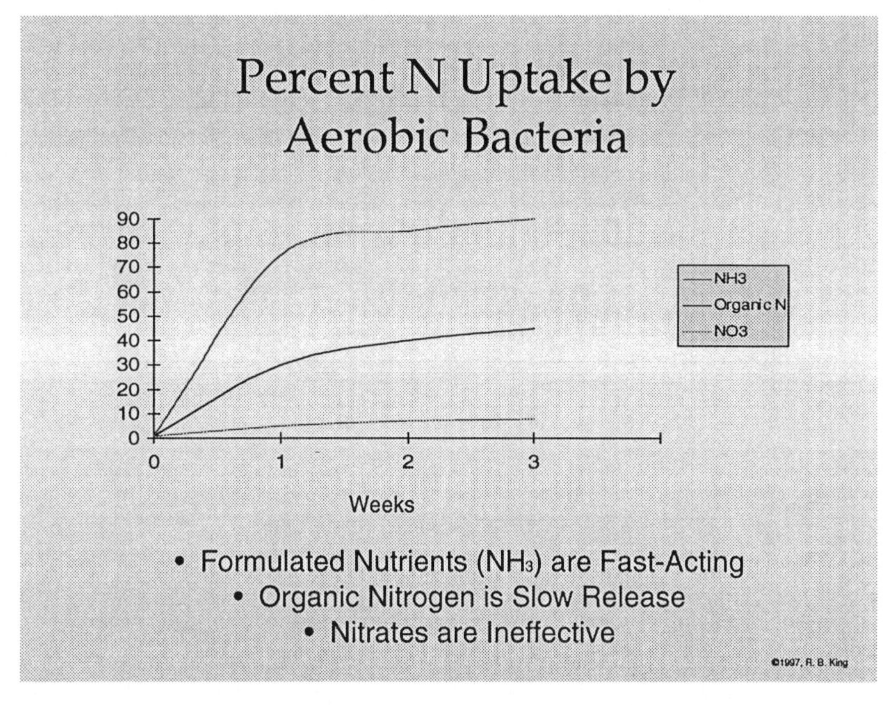

Figure 7-7 Percent N uptake by aerobic bacteria

good aeration by tillage. The nutrient blend specified as a result of treatability testing can be layered into the soils during loading of the treatment cell. A good commercial supplier of bulk nutrients can assist in this phase of the operation.[1]

A few words are in order about nutrients. Land treatment utilizes the services of aerobic soil bacteria that exhibit a preference for nutrient blends high in ammonia nitrogen. Many commercially available fertilizer products are high in total nitrogen, but low in ammonia nitrogen. Nutrient formulations are available that will slow the degradation rate. Some have been observed to stall a project completely. Use of these products will serve to lenghthen the time to closure of a project (see Figure 7-7). Many times a project has stalled and the operators have never found the reason. They simply did not know where or how to look for the problem (which can be detrimental chemicals in the applied nutrient blend). The judicious application of the proper blend of formulated and organic nutrients (low in nitrate nitrogen) can bring a project to closure in a fraction of the time that is likely when the operators are ill-informed about this reality.

The commercial blending of agricultural and garden fertilizers involves addition of ingredients that are unwanted in field bioremediation. Certain chemicals are included in these products by the blendor for preservation against microbial attack. This is done to improve the shelf life of bagged materials and to reduce dusting losses. Antimicrobial additives are common in these trademarked, patented, and proprietary blends. In the practice of field bioremediation for soil treatment, it is prudent to seek the advice and blending expertise of specialists who are knowledgeable in this area in order to avoid the difficulties that are common.[1]

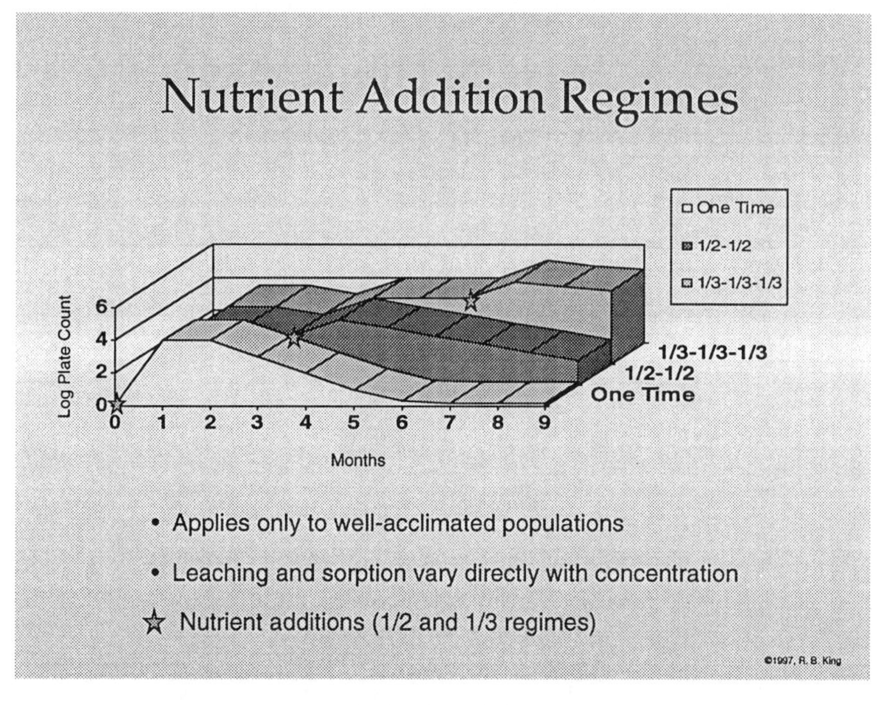

Figure 7-8 Nutrient addition regimes

The periodic addition of nutrients often outperforms the one-time dosing often seen as typical in the field. Many factors combine to vary the way in which nutrients can be most effectively added to the project soils. Figure 7-8 represents data taken from microcosms run on well-acclimated populations. The one-time addition gave good initial growth, but long-term performance suffered as the soluble ammonia nitrogen leached out. When one half the calculated nitrogen is applied initially and is followed by the second half after 3 months, long-term growth is much improved. The best performance over 9 months was observed when one third additions were made initially and at 3 and 6 months. This regime was found to resist leaching by repeated watering over the treatment period.

OPERATIONS AND MONITORING

Once the cell is designed, constructed, and loaded according to the above principles, the operations phase may begin. The soil is normally tilled the day after each significant rain (greater than 1 in. per 24 hr) or after 2 weeks without a significant rain or for covered cells. The soil moisture target is approximately 50 to 80% of field capacity, so water may be pumped and sprinkled from the sump during dry spells if needed. The tilling objective is to provide exposure to atmospheric oxygen throughout the soil pile.

Early in the operations phase of the remediation, the air surrounding the treatment pad should be tested to insure that volatile emissions from the cell surface do not present a health hazard to the surrounding area. The testing protocol may be simply photoionization detector (PID) monitoring or could include whole air sampling and GC analysis for volatile priority pollutants (typically TO14 analysis). For large areas, Fourier transform infrared (FTIR) testing may be used to estimate the emission rates from the cells. If a problem or potential problem is detected, remedial actions need to be taken and the frequency of monitoring will have to be increased. In some cases, sprinkling water over the cell will minimize any volatile emissions. The cell cover will also reduce emissions except during tilling or irrigation.

Soil nutrients, pH, and buffer capacity (lime requirement) tests may be run on soil samples on a monthly basis (or more frequently at the start of operations) to provide an indication of the progress of the remediation and the need for adjustments to keep the process within its target ranges. Accumulated water in the sump may also be tested for these parameters. These tests may be run in the field using agronomic test kits; lab testing should be performed on a periodic (often quarterly) basis to confirm the field results. Microbial plate counts may also be run quarterly to monitor the relative health of the microbial population.

As mentioned in Chapter 2, the target pH range is generally 6.5 to 8.0 for most land treatment units. Typical target values for ammonium nitrogen and phosphate in the soils or sludges are 50 and 5 mg/kg, respectively. The buffer capacity may be expressed as cation exchange capacity or a measure of the limestone required to raise the pH from its existing level to a reference level, typically 7.0. This value will be site-specific initially and will change as the remediation progresses. The objective is to add limestone in increments and approach the desired pH in steps rather than add sufficient limestone to reach the target in one step and potentially shock the soil/sludge system.

The plate counts or MPN results for total heterotrophic bacteria should usually be above 10^6 colony-forming units per gram of soil (CFU/g) after the initial acclimation, and levels of 10^{12} CFU/g have been observed. Once acclimation has taken place, the population of hydrocarbon degraders usually grows to become at least 10% of the total heterotroph population, and may grow to become dominant. For example, if the total heterotrophic population is measured as 10^8 CFU/g, the hydrocarbon degrader population would be expected to be 10^7 to 10^8 CFU/g.

The soil cell is normally divided into grid squares for sampling purposes. The size of the squares is affected by the depth of the soil, the shape of the cell, and the sampling density requirements of the appropriate agency. Some agencies require grid squares to encompass 100 yd^3 of soil or less, and may require compositing techniques for the sample taken to represent each 100 yd^3.

Not all of the squares may need to be sampled during each sampling event. A method where half of the squares are sampled each month is often accepted; in the following month, some of the same squares are sampled and some of the others are sampled. The objective is to provide an indication as to the progress of the remediation.

Based on that reasoning, a relatively inexpensive tracking parameter such as TPH or oil and grease should be considered for the analysis at this point, even if the final closure depends on a specific compound analysis. Many states are using health risk-based closure levels of contaminants in soil, which require that specific compounds be analyzed at a greater cost per analysis. This can be done for all of the squares at the end of the remediation, but is not as cost effective when it is performed from the beginning on all of the squares.

All of these data should be compiled monthly to measure the ratio of carbon to nitrogen to phosphorus (C:N:P) in the soil, the pH and buffer capacity, and moisture content. All of these are used to determine whether nutrients in the form of fertilizer, lime, or water need to be added to the cell. Commercial spreading equipment may be used to add these items if necessary.

Some areas may require that piezometers be installed downgradient from the cell to demonstrate that no contamination is migrating away from the cell. These may require periodic (quarterly) sampling and analysis for target compounds. Monitoring wells downgradient of the cell may need to be sampled even if they were installed for delineation purposes.

A word about decontamination and training is in order. Most sites where this work is performed will be governed by OSHA regulation 29 CFR 1910.120, which means that all workers who may be exposed to the soil are required to complete 40 hr of health and safety training. The neighborhood farmer has probably not completed this training and so may not be hired to till the soil unless he completes the required training (at your expense). Also, the equipment brought onto the cell must be properly decontaminated prior to its leaving the cell. Usually, a thorough wash using water pumped from the sump is sufficient; this is performed on the equipment at the inside edge of the ramp to contain the runoff.

CLOSURE

When the appropriate measure of soil contamination (TPH or specific compound) has been reached, the final sampling and closure steps should be initiated. The final sampling round requires that all grid squares be sampled for the parameter, which is the basis for closure. Increasingly, this is a specific class of compounds such as priority pollutant base neutrals, acid-extractable compounds, BTEX for gasoline-contaminated soils, or specific chemicals where they are the basis for treatment. In some states, TPH may still be used as a basis for closure.

The final sampling protocol usually will include a specific quality control specification for the data and certification of the laboratory performing the work.

When the analytical results indicate that the specified contamination indicators are below the target values, a letter presenting the data and requesting closure is presented to the appropriate agency. If the agency has been following the course of the remediation, it helps to solicit advice from the caseworker as to the form and content of this letter or report.

Once closure has been granted, the berm is removed and the site is restored per an approved closure plan. This may simply require grass seed over the soil, or the soil may be moved to another part of the site for use as fill.

VARIATIONS

The approach outlined above can been modified in various ways to accommodate the needs of a particular site. Two popular options that are briefly described below are the biopile (or biovault) and the biopit.

Biopile

This method can be used (1) where the soil is relatively tight (clay or heavy silt), (2) where excavation must occur for other reasons, (3) where there is insufficient room for landfarming, or (4) where tighter process control is required (e.g., more difficult compounds to degrade). Where insufficient land space is available, the soil may be piled into "lifts" with permeable layers between them. The lifts are typically 2 ft or more in thickness. Slotted piping is installed in the permeable layers, and air is either pulled (using a vacuum pump) or pushed (using a blower) through the permeable zone to introduce air flow through the soil. Nutrients may be blended into the soil before it is placed on the pile, or applied along with the irrigation water that is added by sprinkling or through soaker hoses on the top layer. The typical biopile construction is illustrated in Figure 7-9.

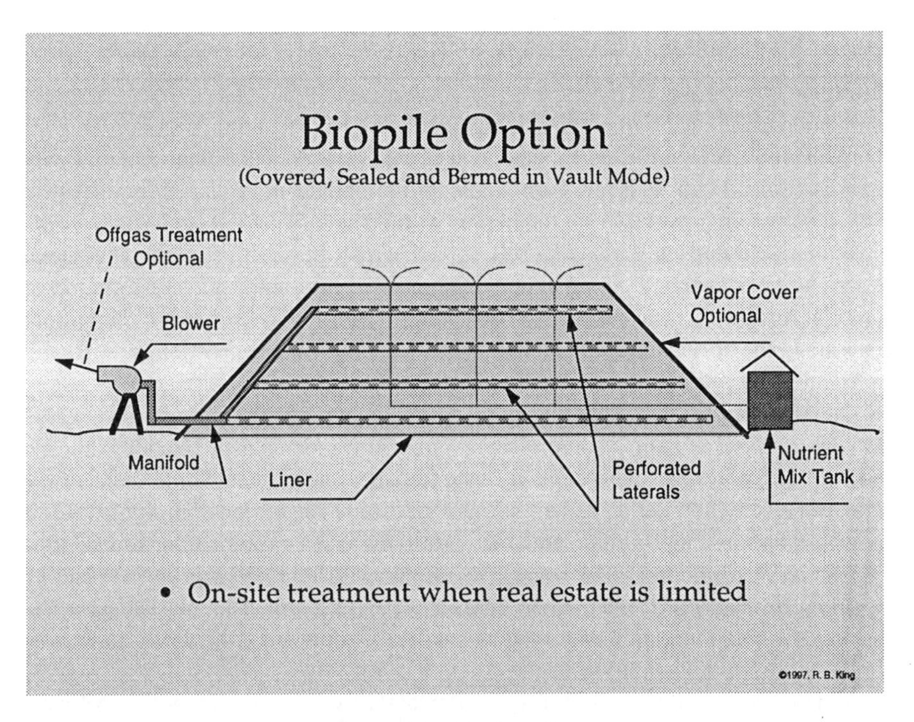

Figure 7-9 Biopile option

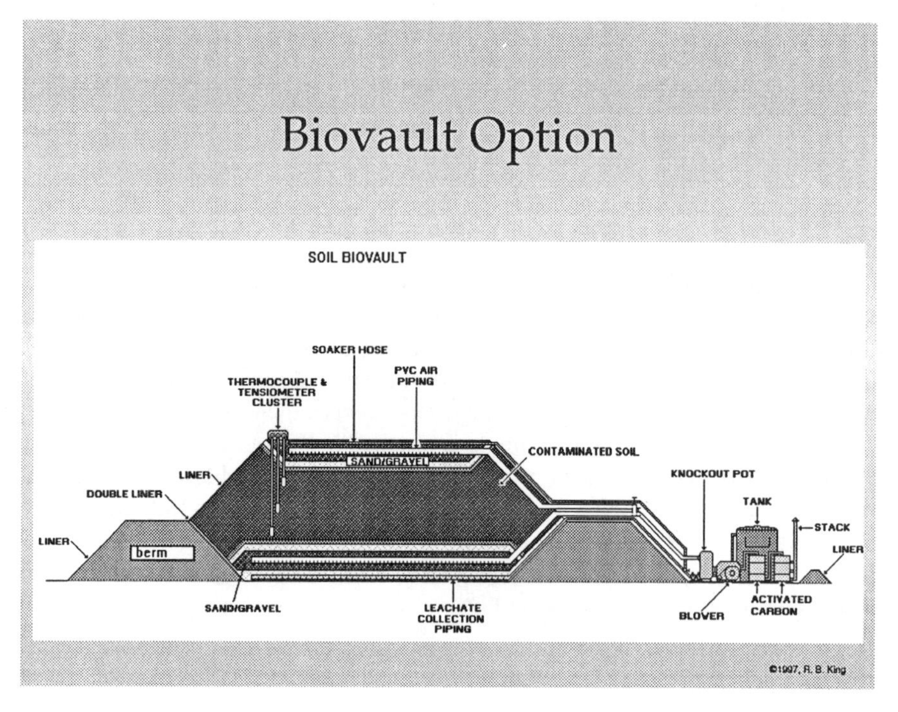

Figure 7-10 Biovault option

Biovault

A biovault is constructed by placing a plastic or other underliner material over a prepared sand bed with low berm walls on the perimeter. An air piping network is then placed in gravel to form the bottom of the vault. A geofelt is placed over the gravel, and the impacted soil is loosely spread across the vault base and piled to the design height. A second piping system is embedded in sand or gravel on the top of the pile. A top cover liner is placed over the pile and may be welded to the bottom liner along the bottom edge to create a sealed bag. Typical construction is illustrated in Figure 7-10.

Air is drawn from the bottom piping system using a vacuum pump (or injected into the top system using a compressor). Water and nutrients may be added through the top piping system or through a separate irrigation system. Some means may need to be provided for air induction into the vault. The water content of the soil may be monitored using commercial irrigation moisture meters and/or probes.

For more recalcitrant compounds where bioavailability is a problem (usually very low solubility in water), dilute concentrations of surfactants may be added to the water/nutrient mix to improve the bioavailability and hence the degradation rate. This is most effective on heavier fuels (such as #4 oil). However, as stated previously, this should not be attempted prior to a biotreatability test for competition between the surfactant and the target compounds for preferential degradation.

A detailed variation on this approach was performed as part of an EPA SITE Demonstration project in New York, in conjunction with the New York Department of Environmental Conservation and the New York Center for Hazardous Waste Management. This demonstration was for two biovaults of 100 yd^3 of soil each, with one vault being operated in a "barely aerobic" mode and the other vault intended to operate in anaerobic mode. The vaults were constructed with a double liner system on the bottom, a single 40-mil HDPE liner on the sides and top, and completely welded seams to minimize any air escape. A sand layer was placed below and above the soil to be treated, with air connections to a vacuum pump from the bottom, a passive air inlet from the top, and a soaker hose arrangement to provide water as needed.

The barely aerobic vault was operated at an exhaust air concentration from a vacuum pump of 5% oxygen; this yielded a carbon dioxide concentration of 12 to 15%. The 5% value was selected because it could reliably be distinguished from 0% oxygen with simple test equipment. The initial air flow rate through the vault was two pore volumes per day, tapering off to approximately two pore volumes every week. The original design was for a lower flow rate toward the end of the test, but the oxygen content approached zero and more frequent testing and air movement were required.

An interesting feature of this test was that six specific volatile compounds were selected to determine how much of the disappearance of the compounds could be attributed to bioremediation: three ketones (acetone, methylethyl ketone, and methylisobutyl ketone) and three chlorinated hydrocarbons (cis-1,2-dichloroethylene, trichloroethylene, and perchloroethylene). Historically, the aerobic degradation of trichloroethylene at significant rates was not well demonstrated at this scale; most of the development work available was done on a bench scale. The test results indicated that over half of the specific compound disappearance for the chlorinated compounds was attributable to biodegradation, with most of the balance being stripped from the soil through the vacuum pump in the first 2 weeks of operation. Virtually all of the disappearance of the ketones was attributed to biodegradation.

Biopit

The same process can be installed in a lined pit below grade and covered with asphalt to render the surface ready for commercial use. Air can be drawn through the soil after nutrients have been added during loading of the soil into the pit. Water can be introduced through the perforated laterals used for air movement. This option is depicted in Figure 7-11.

PROJECT EXERCISE

In order to demonstrate the logic of the design process as described above, we will work step by step through a sample problem from a hypothetical site. While this site is fictitious, the principles and some of the site peculiarities are taken from the authors' experience with actual sites.

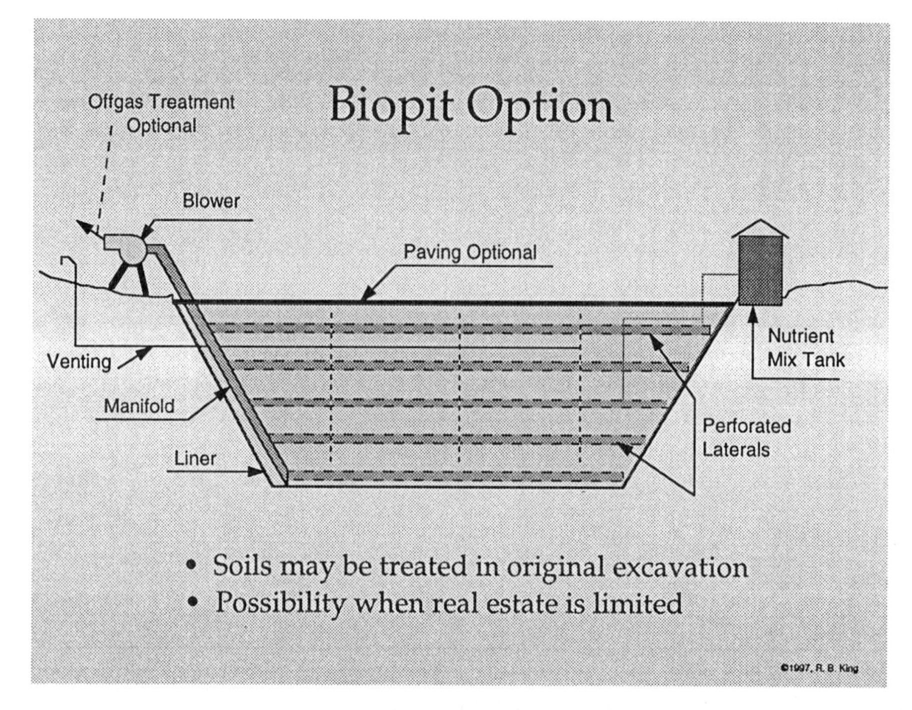

Figure 7-11 Biopit option

A toy manufacturing facility had a loading dock in the back of the plant to handle all the plant shipping. There is a 20,000-gal diesel fuel tank under the dock which was used to fuel the truck fleet. The plant went out of business and the property was put up for sale. The general plant layout is illustrated in Figure 7-12.

During the Phase I Property Transfer Assessment for a prospective buyer, the tank registration issue was raised and the bank required that the seller test the tank and the surrounding soils. The testing discovered that the tank had been leaking from the discharge piping and that the surrounding soils had been impacted. A sampling grid was established as shown in Figure 7-13, and the results of the sampling showed that all of the points shown were above the industrial risk-based action level for the state. The points adjacent to the building were close to the action level, but the points north and east of the tank were more highly impacted. The tank and the contaminated soil both extended 15 ft below the surface, which was paved. A groundwater monitoring well was installed in the downgradient direction to determine whether the fuel had reached the water table; the water table was at 35 ft and the sample showed no diesel components in the groundwater.

The soil area encompassed by the sampling grid was 50 ft out from the building wall and 80 ft long. The local building code prescribed that any excavation adjacent to a load-bearing wall had to approach no more than 8 ft from the foundation and had to be sloped 45° away from the wall.

The sample results averaged 5000 mg/kg total petroleum hydrocarbons (TPH) and a total of 100 mg/kg specific polyaromatic hydrocarbons (PAHs). The state's

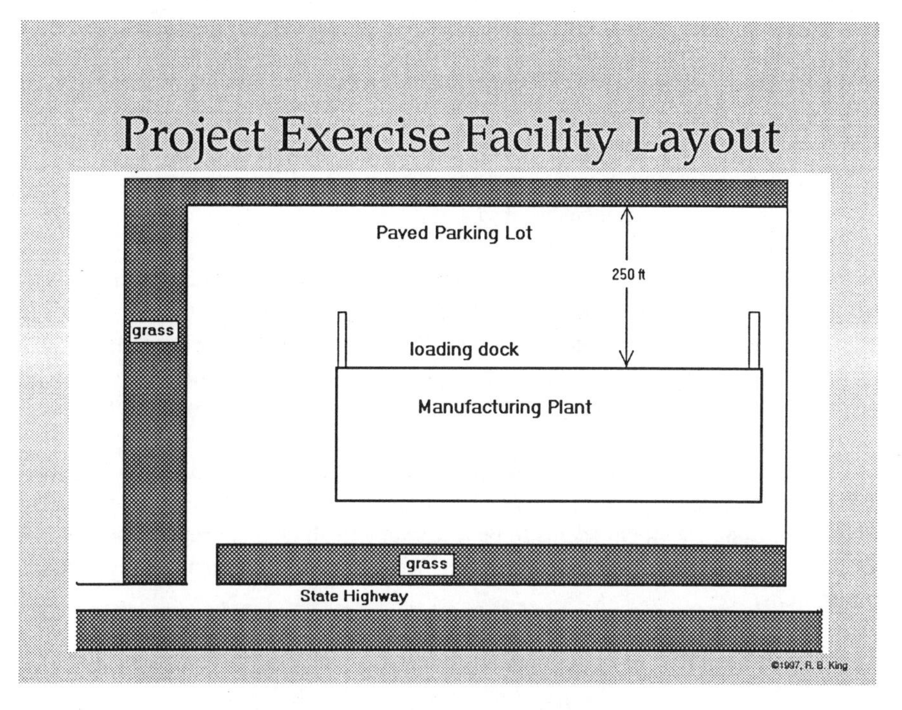

Figure 7-12 Project exercise facility layout

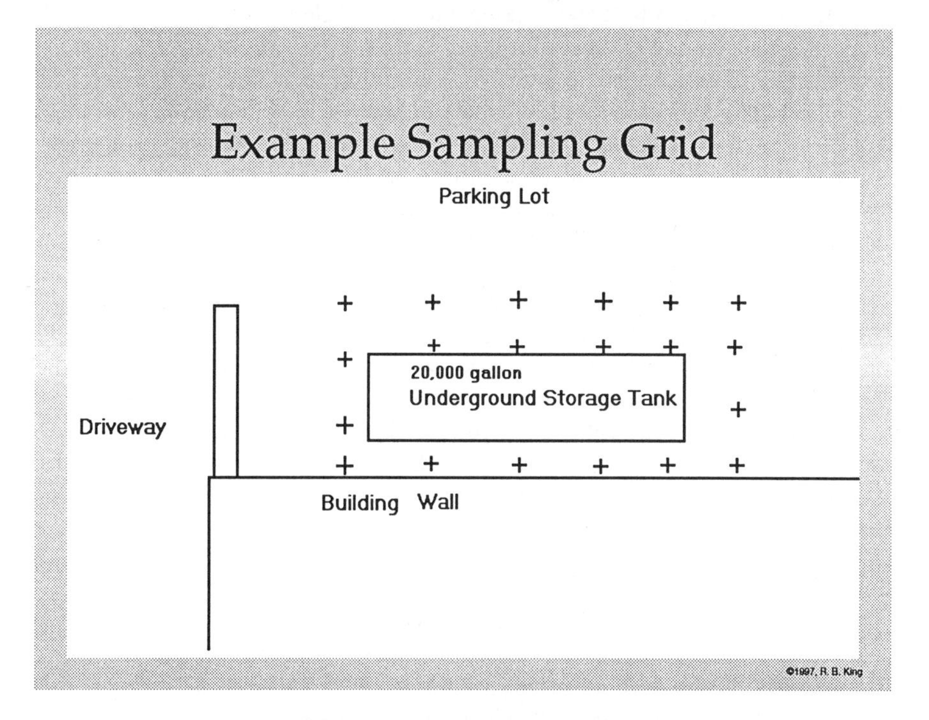

Figure 7-13 Example sampling grid

treatment standard for PAHs is 10 mg/kg total, and the TPH content must be below 10,000 mg/kg.

The property owner wanted a remediation method that met the following criteria:

- No material was to leave the site unless it was absolutely necessary.
- The loading dock had to be restored to use as quickly as possible.
- The building could not be disturbed.

The owner instructed the environmental services company who did the investigation to evaluate ways to remediate the soils based on these criteria. One of the critical questions is "How much money must be put in escrow to cover the remediation so the sale can proceed?" Another critical question is "Can the buyer occupy the premises while a remediation is ongoing with minimal impact on both the buyer and the process?"

Our task is to develop a preliminary cost estimate to answer the escrow question and to develop a conceptual design as a basis for discussion with the buyer regarding occupancy. Land treatment seems to be a workable technical approach, and we will develop our conceptual design based on it. We will proceed by answering a series of questions.

How Much Soil?

The area impacted is 50 × 80 ft, but we cannot excavate up to the building wall. If we start 8 ft from the wall and excavate on a 45° angle from there, we start with an area of 42 × 80 and end up 15 ft down with a 27 × 80-ft hole. Note that the location adjacent to the building prevents us from treating a lot of soil! We assume that the state will permit this soil to remain in place with no additional treatment; we will have removed as much of the source material as possible without compromising the building integrity, and the surface paving will prevent any significant downward migration in the future by preventing any significant rainwater percolation. The loading dock area is likely to always remain paved.

Squaring off the area for averaging purposes, we have $(42 + 27)/2 = 34.5 \times 80 \times 15$ ft, or 41,400 ft^3 (1533 yd^3). If we excavated up to the wall, we would have 60,000 ft^3 (2222 yd^3). Based on the delineation, we will be treating the most highly impacted soil. The tank volume is 20,000 gal, or 2673 ft^3 (99 yd^3). This must be subtracted from the soil volume since it represents void space.

We will excavate 1434 yd^3 of soil, which will fluff when we handle it, so using a factor of 1.3, we will stockpile and spread 1864 yd^3. To fill the hole using similar textured soil, we will need to purchase 1533 in-ground $yd^3 \times 1.3 = 1993$ yd^3 for compaction. We will also need to purchase soil for the berm.

How Big an Area?

If we spread the soil 1 ft thick, we have 1434 $yd^3 \times 27 = 38,718$ ft^3 at a depth 1 ft, or 38,718 ft^2. We need to include a sump in our calculations of treatment cell area: if we anticipate covering the cell and maintaining a volume equivalent to 1 in.

of water over the cell area, this is 38,718/12 = 3227 ft^3 of water. At a depth of 0.5 ft (the depth of sand), this is 6454 ft^2. Thus, the total area needed is 38,718 + 6454 = 45,172 ft^2. This is slightly over one acre. The approximate inside dimensions required would thus be 225 × 200 ft. The sump should be approximately 80 by 80 ft.

Note that if we spread the soil 1.5 ft thick and got maximum utilization of a disk tilling set, the area required would be 38,718/1.5 = 25,812 ft^2. The sump size is 25,812/12 = 2151 ft^3 at 0.5 ft deep, or 4302 ft^2, for a total of 30,114 ft^2. This is approximately 150 × 200 ft.

Materials

The berm length at a 1 ft depth is (2 × 225) + (2 × 200) = 850 linear ft. Since we anticipate that 6 in. of sand will be placed below the soil, and we need 12 in. of freeboard above the soil, the berm must be 30 in. high. We want a 12-in.-wide walkway on the top of the berm, with a 45° slope on both sides. This requires that the berm base be 72 in. wide (30 in. each side plus 12 in. on top). The cross-sectional area of the berm is thus [(12 + 72)/2] × 30/144 = 8.75 ft^2. The volume of compacted soil in the berm needs to be 8.75 × 850 = 7438 ft^3, or 275 yd^3. Since we will buy the soil "loose", we need 275 × 1.3 = 358 yd^3 of soil.

We also need 6 in. of coarse sand for the base. This will be 38,718 ft^2 = 19,359 ft^3 or 717 yd^3. We don't need to put sand in the sump.

We have a choice regarding the interface between the berm and the pavement in the parking lot: we can use clay to butter the inside walls or we can line the whole base with a plastic liner. If we use clay, we need 850 ft × 2/12 × 30/12 = 354 ft^3 = 13 yd^3 of clay. Alternatively, we will need to line the entire area, including the berm walls, or 206 × 231 = 47,586 ft^2.

Watering System

We need to be able to cover the site with 1 in. of water in about 4 hr, for an operator to irrigate the site before tilling in one shift. As calculated above, this is 3227 ft^3 of water, or 24,138 gal. The required pumping rate is thus 24,138/(4 × 60) = 100 gpm (gallons per minute). Note that we can reduce this by half if we want to irrigate over 8 hr instead of 4 hr, but the labor costs will be higher.

We need to spread the water evenly over the entire soil area, which is 38,718 ft^2. Using conventional lawn sprinklers which cover an area 30 × 50 ft, we need 38,718/1500 = 26 sprinklers.

We have some design flexibility here. We don't need to buy one 100 gpm pump and construct an elaborate hose manifold to get the water to all those sprinklers; we can use, say, three pumps at 35 gpm and connect each pump to nine sprinkler heads. This still creates the need for a hose manifold, and it must be relatively simple to assemble and disassemble, because it has to be removed every time the landfarm is tilled. A typical sump pump may meet our need for flow rate, but it may not generate enough discharge pressure to drive the sprinkler heads. A typical sprinkler needs at least 25 pounds per square inch gauge (psig) back pressure to function properly (and cover the rated area), and there will be pressure drop losses through the manifold and hoses. Each hose has to carry 100/26 = 5 gpm, which is not a problem for standard garden hoses. We will select 3/4 in. diameter to minimize the pressure loss through the hoses. If we specify a pump capable of 35 gpm of water at a discharge pressure of 40 psig, we will probably be safe. We need three pumps and three manifolds, constructed of copper or brass pipe and fittings, to permit the attachment of nine garden hoses to each manifold. The sprinklers will function better in the long run if we put a screen on the pump discharge to remove solids prior to driving the sprinkler heads.

How Long Might It Take?

If we look at the sample results, the TPH average value is in the appropriate range, and no remediation would normally be required. The PAH average value needs to be reduced by 90%, from an average 100 mg/kg to less than 10 mg/kg. (NOTE: these are average values and may not represent the time requirements for the most impacted soil! We assume that the stockpiling process will allow us to mix the soils to the average value.) Assuming that there are 6 months per operating year where the average temperature is comfortably above 50°F (10°C), and a half-life for "typical" two- to four-ring PAHs is 6 weeks (this is an arbitrary number, but actual half-lives can be obtained for individual compounds), we need three to four half-life cycles, or 20 to 24 weeks, to obtain 90% reduction. (The reduction percentage sequence based on half-life is 50 to 75 to 87.5 to 93.75%, based on 6 weeks to reduce the starting concentration by half.) This ought to be attainable in one season if the treatment is started at the right time (and the cell construction precedes it by an appropriate amount of time for completion). We should plan to start construction in early March and have the unit loaded and ready for treatment by April 15. This date will be dependent on local climate conditions, site access, and contractor scheduling.

Does the Buyer Need the Space?

We approach the buyer with our conceptual design and ask if we can occupy the east end of the loading dock area for 200 ft from the east end of the building, and he can have free access to the west half of the rear of the building (where the loading dock is shown). Since it is now December, we can probably get approval from the state and install the treatment unit by April 15, so we will be done by next December unless the weather is unusually cold. We will have to determine quickly whether the municipality has to sign the operating permit application, and get on the agenda of a council meeting if the mayor is not empowered to sign such agreements without council approval.

We also tell the buyer that we have to excavate the loading dock area to remove the impacted soil, and that the excavation, backfill, and repaving of the dock area will require 3 weeks of time out of service. After the repaving is complete, the dock area for the west half of the building is his to use. At the completion of the treatment, the soil may be spread across the site, sold, or placed in a nonhazardous landfill. He requests that we convert the treatment cell into a picnic area for his employees when we are done by planting grass on the treatment cell after the remediation is complete. After a discussion about winter grasses and filling the sump with berm soils, we have a framework from which to write a work plan for agency approval and figure out approximately how much all this could cost.

What Will the Conceptual Design Cost?

We have a pretty complete conceptual design, so we need to price the items we have determined we need. The prices given below are highly dependent on geography, local markets, shipping costs, availability of clay and soil, and how far from the time this was written (1996) you are. Given all these factors, we recognize that the estimate is a "ballpark" example. Having said that:

Work Plan and Permits:	
Write, review, and submit work plan (including health and safety, quality assurance, scope and schedule); one meeting with agency	$6,000
Submit applications for air, discharge, and operating permits; attend one council meeting	6,000
Work plan and permits	**$12,000**
Excavation and Backfill:	
Excavate and place 1434 yd^3 @ $10 /$yd^3$	$14,340
Backfill and compact 1993 yd^3 @ $17 /$yd^3$	33,881
Repave 42 × 80 area @ $1.10/$ft^2$	3,696
Sampling and analysis to characterize what remains, 6 samples @ 150 each	900
Dock is ready to use for	**$52,817**
Treatment Cell Construction:	
Berm soil, $20 /$yd^3$ compacted in place × 358	$7,160
Clay, 13 yd^3 × $10 /$yd^3$ at site	130
Labor to spread clay, 8 hr @ $50/hr (including machinery)	400
Sand base, 717 yd^3 @ $12 /$yd^3$ on site	8,604
Top liner, reinforced 8 mil tied outside berm $0.20/$ft^2$ × (210 × 235)	9,870
Cell Preparation	**$26,164**
Equipment:	
3- to 35-gpm pumps @ $800	$ 2,400
Manifolds and hoses	800
Frac tank rental for overflow, 6 months @ $900	5,400
Tractor/disk set purchase, used	5,500
Total water system	**$14,100**
Operating Costs:	
Operating labor, 16 hr/week × 24 weeks @ $25/hr includes tilling, fertilizing, watering, and sampling	$9,600
Miscellaneous test kits and supplies including fertilizer and lime	1,000
Laboratory costs, 8 samples/month @ $150 × 6	7,200
Monthly reports, 6 @ $1200	7,200
Operating cost	**$25,000**

Closure:	
Final sampling, 16 samples @ $150	$ 2,400
Grading and grass planting, $0.15/ft^2	6,750
Final report to state agency	12,000
Closure	**$21,150**

TOTAL COST ESTIMATE **$151,231**

Dividing by 1434 yd^3 treated, we have a unit cost of $105 /yd^3, of which $37,577 ($26 /yd^3) is backfilling and repaving. Our basic unit cost is about $80 /yd^3, which is not unusual for this technology. A cost breakdown per cubic yard is instructive:

Work Plan and Permit	$ 8.37
Excavation and Loading Cell	10.62
Render Loading Dock Usable (not process)	26.20
Cell Construction and Equipment	28.08
Operations	17.43
Closure	14.75
Total	$105.45

Note that the cost to build the treatment cell is about $40 /yd^3 and operations cost about $17 /yd^3. These values are not unusual, and you can see that elimination or addition of construction items can have a substantial impact on the cost. If we had to use a heavy plastic liner under the cell at $1.00/ft^2 installed, the additional cost would be $45,000 or $31 /yd^3 more.

So we suggest to the seller that he needs approximately $150,000 reserve WITH NO CONTINGENCY, and that a project of this sort should have a contingency in the range of 25% to cover weather-related issues over which no one has control. Thus, the impact of the remediation using land treatment on the sale price is effectively about $190,000.

Hopefully this example provides a feel for the relative design issues and the impacts of design decisions on project costs. The unit costs are heavily dependent on the volume of soil to be treated and the number of times the cell can be used; since one of the largest cost components is the cell construction and equipment, reusing the cell and equipment three times will reduce the unit construction costs by a factor of almost three (assuming some small refurbishment is required between loadings). Additional liner systems, backfilling, and other nonprocess-related considerations can also determine whether or not a project is a viable option economically. Some of the cost components are not dependent on the size of the treatment cell (e.g., permits, design, and reporting). While the ranges provided above are useful, each site will have its own peculiarities of design and its own cost structure.

REFERENCES

1. McMillan, D., Desert Green Bioremediation Nutrients and Equipment, Belen, NM, (505)864-2829.

CHAPTER 8

Bioreactors: The Technology of Total Control

INTRODUCTION

In prior chapters we have discussed implementing various approaches to bioremediation where the matrix itself governs the success of the process. This chapter attempts to summarize a broad, competitive, and extremely innovative technology which finds itself quickly moving toward the forefront of commercial bioremediation. No one has yet fabricated the absolute bioreactor — that universal reactor capable of fitting every possible cleanup situation. However, the importance of this approach to applied bioremediation (in terms of total volume of treatable waste) is perhaps unparalleled among the other techniques we have described.

The bioreactor segment of the biotreatment industry is the one area of bioremediation where engineers and scientists can turn the tables on nature and create a near-perfect environment for biodegradation to take place. This technique of "taking the elements from their elements" and, more precisely, manipulating the elements of nature is as close as we can get to a technology of total control. The appeal of this strong sense of control creates an atmosphere of greater understanding and acceptance of the technology between potential end users and regulators. The close ties of this approach to conventional wastewater treatment also aid in its favorable perception, as many of today's environmental professionals began their careers in the areas of wastewater and sanitary engineering.[1]

In the general scope of this chapter, we will define the different aerobic reactor configurations (anaerobic reactors were covered in Chapter 6) and the advantages/disadvantages of each type of reactor. We will look at terminology associated with bioreactors and at some conceptual designs. Finally, we will look at a practical example or case history where a bioreactor was successfully utilized to accomplish a surface water remediation.

PROCESS DEFINITION

Clearly, the comprehensive definition of a bioreactor process contains many factors and this approach will be complicated by a myriad of detail if one is to obtain "total control" over the biological activity within the reactor walls. A bioreactor is nothing more than a reaction vessel, cylindrical or rectangular, that can be stationary, skid mounted, or mobile. This vessel typically has a blower/diffuser system for the provision of oxygen and devices for mixing. It also contains a nutrient delivery system for provision of nitrogen and phosphorus, as well as any other specific nutritional requirements for microbial growth, such as micronutrients or pH control agents.

Bioreactors usually have influent and effluent pumps. Many have the ability to run in a batch or continuous mode. They can often perform treatment via a single pass through the reactor. Other types of reactors require a recycle type of operation. All bioreactors utilize water to provide an aqueous matrix regardless of their treatment purpose.

Other common factors are that they each produce a certain volume of sludge or sloughed biomass; they all produce a certain concentration of off gas which may be CO_2, hydrogen, methane, hydrogen sulfide, or nitrogen (depending upon the contaminants present and the makeup of the microbial consortium contained within the reactor); and all bioreactors have the flexibility of providing primary, secondary, or tertiary treatment.

It would not be unusual to see a smorgasbord of application technologies in use upstream of the bioreactor in a typical treatment train. Among these front-end technologies might be debris removal, particulate filtration, membrane filtration, precipitation, or UV/ozone/peroxide units followed by the reactor. On the effluent end (tail end) of the reactor one might find centrifugation, dewatering presses, carbon adsorption, or air stripping equipment in response to meeting the ultimate remediation criteria at a given site. Figure 8.1 shows a typical bioreactor system and gives examples of possible pretreatment and posttreatment or polishing technologies that might be required to meet regulatory criteria on a specific project.

Although there are several high-tech aspects involved in modern bioreactor design and operation, the overall approach is far from anything representing a "black box". All the aspects of this technology are straightforward. They may not always be simple, but they are definable. The basis for this particular application is to take the best thinking from the fields of chemistry, microbiology, and engineering to create a controlled process in which the biomass within the reactor provides the mechanism of destruction. The best we can do is to understand all the associated factors and design the system accordingly.

BIOREACTOR CONFIGURATIONS

If bioreactors share so much in common, then how could there possibly be very many types of reactors? The several types that we will cover are different principally in their internal engineering. The internal engineering of the reactor is the most important factor accounting for the survival of the many successful bioremediation firms surviving in today's marketplace. It is this internal engineering that has led to the patents and proprietary technologies so cherished in the field of biotechnology.

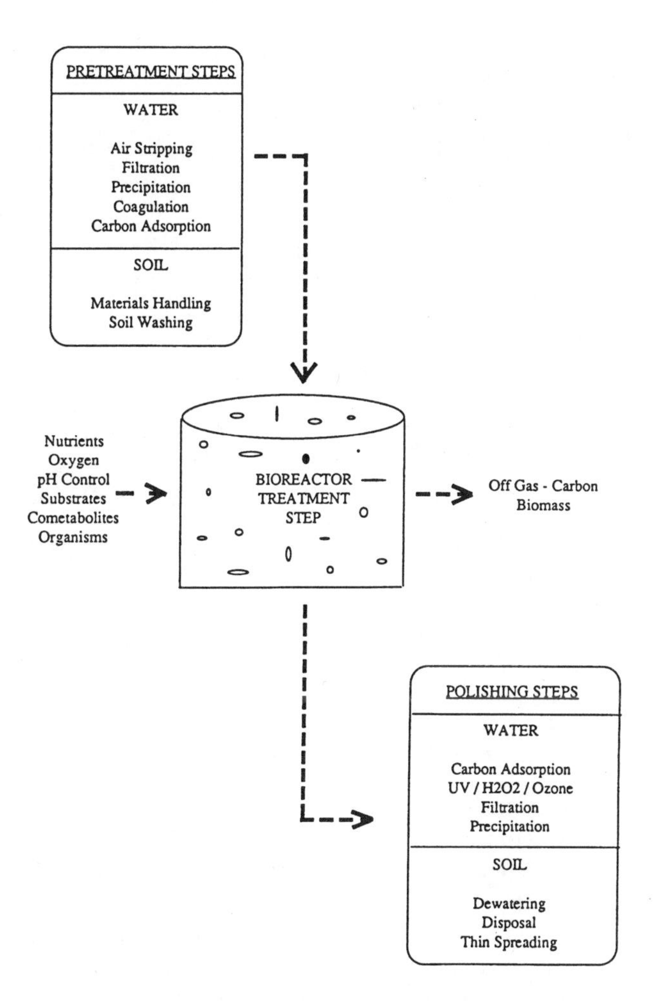

Figure 8-1 A typical bioreactor system

The types of bioreactors are, in fact, numerous. Fixed-film, totally submerged, plug flow, fluidized bed, and sequencing batch reactors are all used in commercial bioremediation. In addition, there are several classic wastewater treatment processes available. These include activated sludge systems, rotating biological contactors, trickling filters, aeration lagoons, and extended aeration. There are soil slurry reactors and reactors that are designed to treat contaminants in the vapor phase (biofilters).

It is important to understand that bioreactor technology is no different than other types of bioremediation in that microbes, contaminants, nutrients, and oxygen must all be brought into contact in a favorable environment. If one thinks of this in the context of the bioremediation triangle as depicted in Figure 8-2, it is clear that each of these three components must come together in a favorable environment at the point of contact in order for there to be a process. If any one of the critical components is missing and there still appears to be contaminant reduction, then there are physical and/or chemical processes at work and the process is clearly NOT bioremediation.

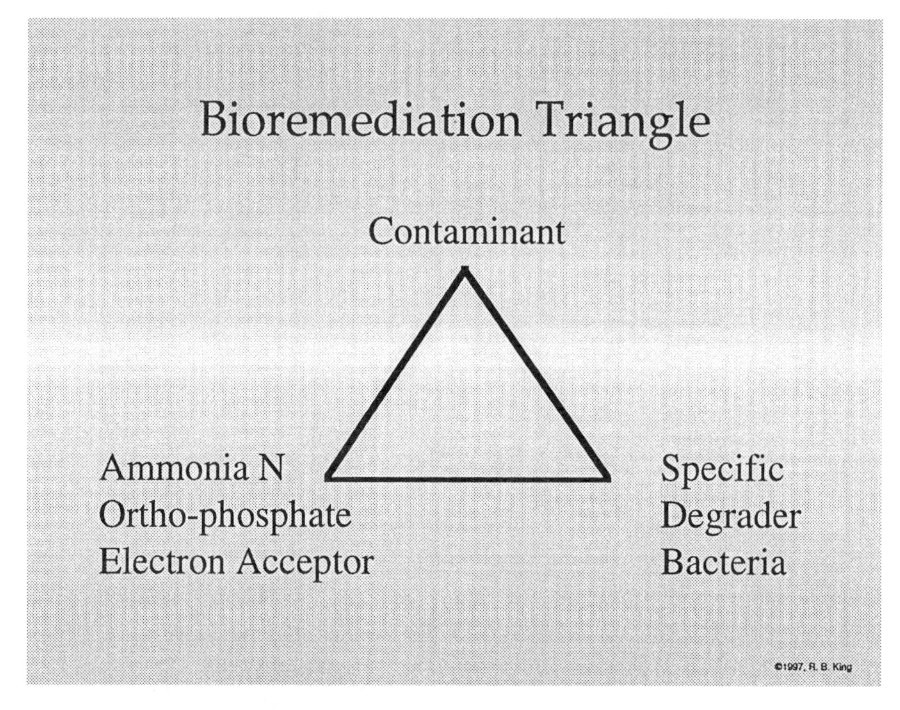

Figure 8-2 Bioremediation triangle

Based on the above theory and the manner in which the three components are brought together, it is interesting to note that bioreactors are classified into two separate categories. Essentially, this is determined by the manner in which they provide contact between the biomass and the contaminants entering the system. These categories are (1) suspended growth and (2) fixed film.[2]

In the suspended growth reactor, all of the biomass is in aqueous suspension, or in the "soup". Contact with the dissolved contaminants is by chance and it becomes a challenging task to maintain a highly functional biomass in the system at any given time. On the other hand, fixed-film reactors have the biomass immobilized and attached to inert surfaces which can be plastic, silica sand, cobbles, or diatomaceous earth. In this case, the dissolved contaminants pass over the microbes and the biomass remains held in the system for long periods of time.

Table 8-1 gives the pros and cons of a variety of bioreactors. The following is a short description of many of the reactors along with a more elaborate description of the types of reactors that are gaining considerable attention: fixed-film, submerged plug flow, and soil slurry reactors.

Activated Sludge

This process typically involves a two-phased approach to treatment of process wastewater, surface water, or groundwater. The technique employs the suspension of microbes in an aqueous waste-filled tank. Nutrients are added and the tank is

Table 8-1 Bioreactor Pros and Cons

Bioreactor type	Pros	Cons
Activated sludge	Treats high BOD, COD, TOC	High cost
	Long history of usage	Labor intensive
	Excellent pretreatment	Produces high sludge yield
Sequencing batch	Handles influent fluctuations	Requires strict solids control
	High reliability	Temperature sensitive
Rotating biological contactor	Shock resistant	Poor vapor emissions control
	Supports lush biomass	Inconsistent treatment
Trickling filter	Short retention times	Poor shock response
	Minimal vapor emissions	Prone to dead spots
Soil slurry	Short retention times	High cost
	Excellent control	Labor intensive
Biofilter	Handles high flow rates	Short history of usage
	Excellent economics	Short usable life
Fixed film	Low sludge production	Requires vapor, iron, solids control
	No sludge recycle required	Marginal performance at low contaminant levels
	Shock resistant	

aerated. All treatment takes place in this tank and a clarification step for removal of excess biomass follows. This involves the settling of the biological sludge, some of which is recycled back to the front end of the plant, and the remainder is wasted. Finally, treated effluent is discharged from the clarifier tank skimmers.

Activated sludge systems have found wide applicability in situations where there are high BOD, TOC, or COD concentrations in the wastewater.[2] It is not uncommon to see activated sludge used as a pretreatment step prior to a fixed-film bioreactor. They are often used to knock down gross concentrations of organic carbon. The chief disadvantages have been their relatively high cost, their inability to deal with variations in influent concentrations, high costs due to labor requirements, difficulty in controlling volatile emissions, and the need to deal with large volumes of waste sludge which may contain adsorbed contaminants.

Sequencing Batch Reactors (SBRs)

SBR technology is one of the more highly useful and versatile reactor approaches. While having many similarities to the activated sludge system described above, it differs in that the entire operation is conducted in one tank. The typical mode of operation is first to acclimate a population of microbes contained in the treatment tank prior to their contact with the waste. After acclimation is complete, waste is introduced. On a timed basis, the contents of the tank are thoroughly mixed and aerated to maximize biodegradation.

The next step in the process involves a complete stoppage of all activity within the tank and the solids are allowed to settle. This is followed by a step where the liquid is drawn off and further polished (or is sent for discharge). Finally, a portion of the acclimated sludge is retained in the tank for the next round of treatment. SBR technology is ideal for batch-type waste streams and has greater tolerance to fluctuations in influent contaminant concentrations.[3] It has proven to be an extremely reliable approach, but it does have some disadvantages. Among the disadvantages are the temperature sensitivity of the system, the need for rigorous solids control, and the need to provide treatment for vapor emissions. The hydraulic retention time in the tank can be rather substantial.

Rotating Biological Contactors (RBCs)

RBC technology has characteristics of both fixed-film and activated sludge systems. In this approach, polyethylene disks are mounted in series on a turning shaft and the assembly is mounted in a tank half filled with wastewater. Biomass grows as a mat on the disks as they rotate through the waste. As much as 40% of the disk surface can be submerged in the waste at any given time. When the disk is not submerged, it is being exposed to oxygen to help drive the aerobic biological reactions. Excess biomass sloughs off the disks via the shear forces associated with the turning action in the tank.

This type of system tends to support lush biomass and is relatively shock resistant. Interestingly enough, RBCs are not considered able to provide the same level of treatment efficiency as that of an activated sludge system.[4] The system also tends to have significant requirements for sludge disposal and off-gas treatment.

Trickling Filters

Trickling filter systems are closely related to fixed-film reactors to be described under the next subheading, except that they operate in continuous flow mode. They are typically filled with packing material of plastic, cobbles, or with silicate sand, providing a fixed surface to which microbes can attach. The waste slowly passes down over the biomass attached to the packing.

Trickling filters are characterized by very short retention times and must run in a recirculation mode in order to produce significant reductions in contaminant concentrations. These units are also quite adaptable to fluctuating waste streams and have proven to be particularly effective when used in the pretreatment mode. Release of volatiles from these units is not as significant an issue as with other types of reactors described in this chapter.

Disadvantages include difficulty in rebounding if the operation of the system is disrupted for any reason, and possible dead spots that can occur within the unit when proper contact of biomass, contaminants, oxygen, and nutrients fails.

Soil Slurry Reactors

There is a great misconception in the marketplace that there has been a great amount of recent soil slurry reactor project activity. In actuality, there has been a limited amount of activity using this approach and there have been very few field projects of this type.[5] Even fewer projects have been completed without ultimate assistance from other treatment processes. In the future, we will see increasing activity as capital cost for soil slurry equipment decreases and the Land Ban restrictions exert a greater influence on wastes requiring treatment.

The size of currently available full-scale slurry bioreactor units can range in diameter from 10 to 50 ft and from 15 to 25 ft in height. These reactors are sized to hold slurry volumes from 15,000 to 300,000 gal. A general price range for capital equipment, design, and construction (based on the above dimensions) can be on the order of $125,000 to $2 million.

Prior to entry into slurry reactors, soil must be pretreated. This can involve one or more materials handling and preparation steps involving many techniques. It is absolutely essential that the soil which enters the slurry reactor be sized and graded to smaller than 60 mesh so that an adequate slurry can be maintained against gravity. Obviously, silts and clays are better candidates for a slurry reactor as opposed to sands and gravels. Once the materials handling and sizing are completed, soils can be thickened or pretreated with biodegradable nonionic or anionic surfactants prior to loading into the reactor.

One of the more interesting pretreatment techniques before soils are loaded to a slurry reactor involves the use of a commercially available soil washing process. Although this process has had minimal practical application in the U.S., there is good reason to believe that the combination of soil washing and slurry bioreactors will be a likely process for future development. One soil washing process (utilized by BioTrol, Inc. of Chaska, MN) uses technology extensively adapted from the mining industry.[6] Techniques included in this process are froth flotation, attrition scrubbing, vibrating screens, mixing trommels, hydrocyclones, and pug mills coupled with a countercurrent flow of washing fluid (usually water). The soil washing process removes surficial contamination and aggressively scrubs soils so that strongly adsorbed contaminants are released. In doing so, volume reduction of the soil is achieved and there is a separation that takes place in which washed coarse particles (sands) are removed and fines exit the system to a slurry reactor.

Commercially available soil slurry reactors consist of tanks (most typically one to three tanks in series) that provide a combination of aeration and mixing, maximizing contact of the (20 to 40% solids) slurry with indigenous microbes from the soil. Mixing can be accomplished by mechanical means through innovative processes (like that marketed by EIMCO of Salt Lake City, UT), or simply by the action of the aeration mechanism in the tank.

Oxygen, nutrients, chemicals for pH control, and possibly microbial cultures are added. The water used can be tap water if available, or water that is trucked to the facility, groundwater, or surface water. As treatment is completed over a period of several days of retention time within the reactor, the soils are dewatered using centrifugation or a belt filter press. Depending on regulatory approvals, the soils can be reused onsite as fill material, landfilled as nonhazardous waste, or thinspread in road construction fill. Water from the dewatering step is split with half going back to the system and the remainder discharged or passed on for further treatment (as in a publicly owned treatment works).

It is possible for these reactors to run in an anaerobic mode, but most of the experience gained has been in aerobic operation. These reactors offer greater control over environmental conditions than do other types of *ex situ* soil treatments. When properly designed and operated, such things as pH, temperature, and vapor emissions can be fully controlled.

Biodegradation within the reactor proceeds at a rapid rate and offers significant advantages over other commonly employed *ex situ* techniques such as land treatment, land farming, and aerobic soil piles. Of course, as with other types of bioreactors, the biokinetics depend upon the nature of the contaminant (chemical classification and structure), the concentration of the contaminant, and the regulated treatment level that must be attained. Typical treatment times in a slurry bioreactor range from less than 1 month to more than 6 months.

Vapor Phase Bioreactors (Biofilters)

Another area of bioreactor technology receiving a significant amount of development time and money is the vapor phase biofilter.[7] Biofilters are commonly constructed in a vessel packed with loose beds of solid material, soil, or compressed "cakes" with microbes attached to their surfaces. Waste gases are passed through these units via induced or forced draft. The organic content of the vapor is reduced through microbial digestion. They are capable of handling rapid air flow rates and volatile organic carbon (VOC) concentrations in excess of 1000 ppm. These units are gaining in usage and are timely in that they are a cost-effective means by which to deal with the more stringently regulated VOC emission levels.

Since biofilters compete with incineration and carbon adsorption in many situations, they are attractive in terms of not having to deal with landfilling costs or regeneration headaches. This has already been recognized in Europe and some biofilter technology has found its way to the U.S. Also, the thought of not simply transferring contaminants from one medium to another is particularly appealing. The biofilter creates a truly destructive process.

There are essentially two types of biofilters. The first and simplest is the soil filter. Contaminated air from a small waste stream or other treatment process (such as air from a soil vapor extraction process) is passed through a soil compost-type design. Nutrients are preblended into the compost pile to provide conditions for biodegradation of the waste by indigenous bacteria.

The second type of biofilter (and certainly the up-and-coming version) is the treatment bed or disk. In the treatment bed, the waste air stream and the filter are humidified as the waste is passed through one, two, or more beds made up of compost, municipal waste, sand, or diatomaceous earth. In the disk approach, a series of humidified, compressed disks are placed inside a reactor shell. These layered disks contain activated charcoal, nutrients, microbial cultures, and compost material. The waste air stream organics undergo biodegradation as they pass through the disk system. Any collected water condensate from the process is returned to the humidification system for reuse.

Biofilters have reportedly been built to handle up to 90,000 ft^3 per minute (cfm) of air flow using filters up to 20,000 ft^2 in wetted area. These filters can also be customized with specific carriers, nutrient blends, or microbial cultures. Some biofilters can endure up to 5 years before replacement is necessary.[8] Spent filters can be utilized as fertilizer since they present no hazard.

Fixed Film, Plug Flow

The most widely accepted reactor available today is the fixed-film, plug flow unit. This particular reactor has the reputation of being the most effective of all the other types and tends to be so versatile in its treatment ability that it can be used to treat groundwater, surface water, or process water. It frequently can be installed as a "package plant" to treat industrial wastewater. This reactor tends to be somewhat shock-proof; that is, the unit can withstand great fluctuations in organic loadings. Its resident biomass maintains itself for long periods of time.

One of the unique features of these reactors as currently marketed is their design flexibility. A fixed-film, plug flow reactor can be designed with single or multiple cells to enhance the system shock resistance even more. Inside is a submerged packing material (structured or random) which encourages the formation of a microbiological film, which ultimately builds into a lush mat of biomass. With this type of design, sloughing of the biomass is minimized. This leads to an effluent that has only moderate total suspended solids (TSS). The effluent can be discharged if the minor concentrations of biomass floc are of no concern to regulators.

Another key aspect of this type of bioreactor is the fact that it can be sometimes successfully utilized to deal with relatively low levels of organics found in groundwater, and still be able to deal with levels of organics in wastewater often greater than 1000 ppm. This assumes, of course, that a TOC:N:P ratio of 100:5:1 can be maintained. Flow rates through this type of reactor can exceed 100 gpm with a packed volume within the reactor of greater than 1500 ft^3. Commercial fixed-film, plug flow reactors can be as large as 30 ft in length, 9 ft wide, and 10 ft in height. This depends on design flow rates and organic loadings.

If flow rate and organic loading dictate, bioreactors can be staged in series. Also, if several different waste streams require treatment simultaneously, then one could operate these bioreactors in parallel. More and more complex waste streams are destined to treatment via biological systems. These waste streams may have BOD values greater than 5000 ppm and COD values greater than 15,000 ppm. In order to limit the overall size of the bioreactor system, as many as ten fixed-film reactors can be aligned for this type of application.

DESIGN CONSIDERATIONS

As many of the principles of microbiology, chemistry, and geochemistry apply to a system designed to treat groundwater, a water quality study may need to be performed (see Table 8-2).[9] This study can proceed parallel to the screening and treatability studies described in other chapters. This places strong emphasis on key site water quality parameters and their effect on specific equipment requirements. After all, a bioreactor is only as effective as its response to the influent chemistry, and how well the internal environment (and hence the degradation process) of the reactor can be controlled.

Table 8-2 Water/Wastewater Analyses for Bioreactor Applications

Chemical	Microbiological
TSS, COD, TOC, pH	Total heterotrophs, BOD
Metals scan	Specific degraders
N, P, Fe,	Shake flask studies
other specific parameters	Microtox™ assays

- Produces data for short-term, cost-effective results
- Provides key design information and Go/No Go decision
- Identifies potential problem factors

A typical water quality study will involve a standard measurement of pH, specific degradation capabilities of the indigenous microbes in the waste stream, and the level of nutrients available. Also, a metals scan should be conducted to identify any toxicity problems and to anticipate metals that may be absorbed into the biomass. In addition, there are other parameters that may have a major impact on the design of the reactor and equipment requirements that are needed to meet the regulatory standards. These parameters are TSS, iron, BOD, COD, TOC, and the rate of degradation of the contaminants as measured through a column test. Failure to address these parameters can result in an undersized reactor and, consequently, an expensive error. The total duration of all the tests comprising a water quality study with a column study is 6 to 12 weeks. Total costs range from $8000 to $25,000.

Total suspended solids is an important parameter in that their presence in levels greater than 100 ppm can have a dramatic plugging effect within the reactor. Solids present at this level prompt the need for pretreatment filtration or flocculation.

Iron concentrations at greater than 10 ppm soluble iron require pretreatment by precipitation and filtration. Otherwise, the iron could precipitate with the nutrients in the reactor. Residual iron in its oxidized state is likely to leave the system via absorption in the sludge.

The relationship between TOC, BOD, and COD is quite important despite common criticism of the methodologies used to measure these parameters (particularly BOD). Both TOC and COD give a gross indication of the amount of organic carbon available to a microbial population. These concentrations always exceed the BOD. The nutrient balance (or operating concentrations) for a bioreactor can be based on the higher levels of COD or TOC. By doing this, the microbes will never be limited by carbon.

The rate of biodegradation of specific chemical compounds in the wastewater is of particular value in the design of a bioreactor. Typically, packed column studies that will generate this information are used for process design and fine-tuning (see Figure 8-3). These column test designs should be based on the most recent analytical data available.

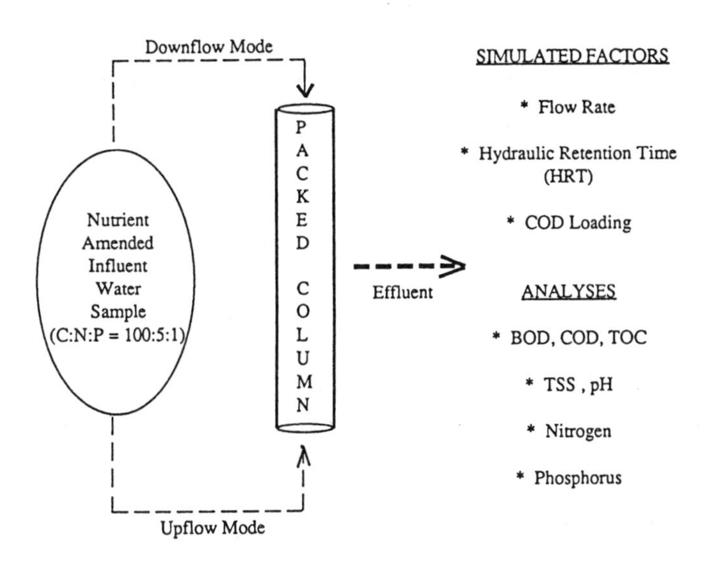

Figure 8-3 Process fine-tuning in the lab

Typically, column tests are conducted in glass vessels that are 3 ft in length and 4 in. in diameter. Air is supplied through an air stone at the base of each column and the columns are packed with plastic saddles that allow the attachment of biomass. Often, columns are foil wrapped to prevent contaminant losses due to photodegradation. Column tests are often run in a batch mode with an initial period of acclimation allotted after a spike of activated sludge seed. During this time period, an attenuated stream of the waste is passed through the column (usually a 25 to 50% dilution of the raw wastewater). The appropriate nutrient additions are made to the water to be treated prior to its entry into the column. Again, a COD:N:P ratio of 100:5:1 is necessary to optimize biodegradation.

The column study protocol usually involves two to three simulations of COD loading (expressed in lb/1000 ft^3/day), hydraulic retention time (HRT), and flow rate. The amount of water required for the entire test is dictated by the number of simulations to be made. At varying time intervals during the test, COD, BOD, TSS, pH, and other critical analytical parameters are measured. When completed, the column test provides the necessary sizing information for the bioreactor and provides indication of other equipment or treatment needs. Also, the biomass yield can be calculated to provide an indication of sludge stability and possible routes of disposal. Results of a column test are often used by bioremediation firms to provide performance guarantees.

PILOT TESTING

In addition to (or in lieu of) the water quality study, pilot studies can be conducted on site using technology and equipment that are identical to that which will ultimately be implemented. The benefit of a pilot study is that the technology can be demonstrated right at the site in an actual field scenario without having to simulate conditions or add new variables. Pilot studies of bioreactor technology are quite common, especially since more and more mobile technology is becoming available.

The typical pilot study lasts 4 to 6 weeks and costs from $15,000 to $50,000. The information gathered can vary greatly, but usually the optimal flow rate and loading can be determined. Often, the levels of nutrients required and the general period of acclimation of the biomass are also determined.

INSTALLATION AND START-UP

The actual design and engineering for a bioreactor can take up to 1 month to complete. Permitting is a state-by-state issue, but the general ability to rapidly permit a bioreactor exists even in states that have had little to no biological treatment activity. When a full-scale bioreactor unit is slated for the field, typical fabrication and delivery times range from 6 to 15 weeks, based on size and complexity of design. When the bioreactor arrives on site, the blowers and diffusers, the nutrient, equalization, and tempering tanks, off-gas treatment, primary treatment cells, pumps, piping, and control panels are assembled. Then comes the need for connection to the water source, identification of the routing of reactor effluent, and preparations for sludge storage or disposal. In addition, the following hints may be helpful during the 2 to 3 days of installation time for a bioreactor:

- All piping that feeds the influent end of the bioreactor or that exits the effluent end needs to be heat traced or buried below the frost line, if freezing conditions are possible.
- A common way to protect the entire treatment system is to place everything within an enclosure.
- As with other types of treatment systems, quick-connecting devices work nicely from a maintenance standpoint and are particularly useful if the system needs to be moved to an alternate location on the site.

Once the bioreactor is installed, the initial flow of water is started and the system goes through a period of equilibration. This can last from 1 to 3 weeks based on the type of system and the nature of the contaminants and biomass. It is quite possible that the bioreactor may discharge directly to a sewer or treatment facility on site before permits are secured for utilization of alternate disposal routes.

As the operation of the bioreactor continues, a monitoring and sampling protocol is carried out both initially and for the entire duration of the project. Sampling and analyses typically occur monthly and involve the measurement of specific parameters like COD, specific contaminant levels, pH, suspended solids, levels of nutrients, and the presence and number of specific degrader organisms within the biomass. Many of these measurements are made on both the influent and the effluent. Off-gas analysis may be made from samples collected at the vent on the reactor. Usually shrouded Tedlar bags are utilized and their contents analyzed via gas chromatography with flame ionization detection (GC/FID).

Table 8-3 Lagoon Water Treatment Case History

Parameter	Influent	Pretreatment	NPDES permit	Average effluent
COD	6700	1900	—	540
BOD	1600	1300	240	115
Pentachlorophenol	36	29	—	0.1
Phenol	7.6	1–96	0.8	0.3
Total PNA	37	1.2–49	—	0.3
Oil and grease	290	5–36	30	8
TSS	2450	580	20	75
pH	7.2	3.8–6.8	6.0–9.0	7.3
Total K-001	52	7.5–104	3.0	1.89

CASE HISTORY[10]

The following case history represents a successful treatment of lagoon water at a site in the Southeast. A complete treatment train utilizing a fixed-film bioreactor accomplished the cleanup. This site was an inactive wood-treating plant whose primary contaminants were pentachlorophenol and creosote. After an initial site assessment was made, a two-phased remediation program was proposed and approved for the site. This program involved treatment of the lagoon water and final discharge to a local river under an NPDES permit. This was followed by the dewatering of the lagoon sludge with placement in an on-site landfill.

Bioremediation was merely one of the proposed methods for the lagoon water cleanup. Other treatment options were dissolved air flotation (DAF), dissolved air flotation plus carbon adsorption, and dissolved air flotation plus chlorine dioxide destruction. All except the bioremediation proved unacceptable because of the inability of the other avenues to attain the levels of BOD and (RCRA) K001 reduction necessary to satisfy the NPDES limits. Also, the timing of the project was critical from a mobilization and a completion standpoint. Bioremediation filled the bill on all three counts.

Table 8-3 shows the initial influent levels of contaminants coming straight from the lagoon. Total COD was 6700 ppm and total BOD was 1600 ppm. Also, total pentachlorophenols at 36 ppm and total PNAs at 37 ppm were significant as was 52 ppm of K001 solvents. This would be a challenging cleanup for any treatment technology. Biological treatment was the method of choice because a system could rapidly be brought to the site, installed, started up, and the treatment could be completed within several months.

After a water quality study demonstrated up front that the contaminants of interest could be degraded through biological treatment in a timely manner, the design for a two-stage treatment system was completed. The system involved pretreatment of the lagoon water by flocculation and an activated sludge system. This was followed by a series of two fixed-film bioreactors. Clarifiers were located (1) at the effluent end of the activated sludge system to capture biomass and for sludge recirculation, and (2) at the end of the fixed-film reactors to reduce suspended solids in the effluent. The system was designed to handle a minimum flow rate of 50 gpm and a COD loading of 200 lb COD/1000 ft^3/day.

The design, construction, and mobilization of the fixed-film units were completed in 8 weeks while the activated sludge and flocculation units were being built on site. Once on line, the activated sludge and flocculation system served to reduce the gross concentrations of COD, BOD, and oil and grease present in the lagoon water. Following that, the fixed-film reactors had the task of degrading all the remaining contaminants to meet the NPDES limits. A specialized culture of bacteria described by Crawford et al.[11] was added to the bioreactors to hasten the biodegradation of total pentachlorophenols.

The average effluent results of this 2.5-month project are presented in Table 8-3. A volume of 2.7 million gallons of water was ultimately treated at an average flow rate of 70 gpm. In an efficient fashion, the system served to reduce BOD to 115 ppm, COD to 540 ppm, and to reduce pentachlorophenol, PNAs, and phenols each to 0.5 ppm or less. The only limit not met by the system was for total suspended solids. However, the superlative overall performance of the system allowed room for successful negotiation of a higher TSS limit.

This project served to establish the efficiency of a fixed-film bioreactor in meeting some aggressive standards for several challenging contaminants at this particular site. More importantly, this case history points out the utility of combining technological approaches to get a particular job done. The object here is always to get the project completed in a timely fashion using the most cost-effective solution that meets the project requirements.

CONCLUSION

This chapter has briefly covered the basic information regarding bioreactor treatment technology. It is obvious that these controllable treatment units have wide applications at present and will most likely be highly utilized in the future. More and more companies are adding bioreactors to their treatment repertoire, while some firms are allocating major time and money expenditures on new designs that will greatly enhance their utility. Over the next few years we will likely see bioreactor systems developed that can routinely treat chlorinated solvents, PCBs, and pesticides.[12] The bioreactor technology of total control will undoubtedly figure significantly in the future of waste bioremediation and reclamation of contaminated sites.

REFERENCES

1. Snyder, J.D., 1990, How biotreatment works, *Environ. Today,* 1(1), 20.
2. U.S. Environmental Protection Agency, 1989, Bioremediation of Hazardous Waste Sites Workshop, CERI-89-11, Washington, D.C.
3. Lewandowski, G.A., 1990, Batch biodegradation of industrial organic compounds using mixed liquor from different POTWs, *Res. J. WPCF,* 62(6), 803–809.
4. Grady, C.P.L., Jr. and Lin, H.C., 1980, *Biological Wastewater Treatment*, Marcel Dekker, New York, pp. 833–876.

5. Ross, D., 1991, Slurry-phase remediation: case studies and cost comparisons, *Remediation*, 1(1), 61–74.
6. Valine, S.C., Chilcote, D.D., and Chresand, T.J., 1990, BioTrol Soil Washing System, presented at the U.S. EPA Second Forum on Innovative Hazardous Waste Treatment Technologies: Domestic and International, Philadelphia, PA.
7. RMT Network, 1991, Air Pollution Control May Be Reduced with Biotechnology, Madison, WI, 6(1), 5–8.
8. Holusha, J., 1991, Using Bacteria to Control Pollution, *New York Times,* Section C6, March, 13.
9. Bourquin, A.W., 1989, Bioremediation of hazardous wastes, *J. Hazardous Mater. Control,* September–October.
10. Wall, W.T. and Chresand, T.J., 1989, unpublished site summary.
11. Crawford, R.L., Brown, E.J., Pignatello, J.J., and Martinson, M.M., 1986, Pentachlorophenol degradation: a pure bacterial culture and an epilithic microbial consortium, *Appl. Environ. Microbiol.,* 52(1), 92–97.
12. Bernstein, K., 1990, Do microbes hold the key to toxic waste cleanup? The EPA thinks so..., *BioWorld,* November, p. 46–51.

CHAPTER 9

In Situ Aquifer Bioremediation

In situ bioremediation encompasses several types of processes, all with the same aim: to utilize microorganisms to degrade hazardous constituents in aquifer soils and groundwater with minimal soil disturbance. Usually the objective is to operate the process in an aerobic mode. This chapter will consider the "classical" approach to *in situ* bioremediation.

HYDROGEOLOGIC VARIABLES

For the purposes of discussion, our basis of reference will be the terms depicted in Figure 9-1. The vadose zone is the unsaturated subsoil environment reaching from natural ground surface down to the capillary fringe where groundwater seeps up into the base of the vadose zone. The capillary fringe varies in thickness depending on the soil makeup and particle size (lithography). The water table is the point of 100% water saturation, lies at the base of the capillary fringe, and is the uppermost extent of the phreatic (saturated) zone. The water table can fluctuate seasonally due to rainfall and the pumping of wells for irrigation and domestic water supplies, as well as pumping during an aquifer remediation. Pumping will also affect the direction of flow (gradient) and the flow rate of the aquifer water. The permeability of soils (both above and below the water table) is the rate at which a fluid can flow through them. Soils of high permeability allow faster flow in centimeters/second than soils of low permeability. The floor of an aquifer is referred to as an aquatard. If this impermeable layer forms the roof of a deeper aquifer, the lower aquifer is said to be confined. If the aquatard is fractured (either naturally or by drilling activities), there can be communication or water flow between these aquifers. Where a surface aquifer is contaminated, but the lower confined aquifer is not, extreme care must be exercised to avoid drilling through the aquatard. Sometimes a clay lens in the vadose zone may trap percolating contaminants and water. This small pocket of contamination must be carefully removed to avoid contamination of the aquifer below.

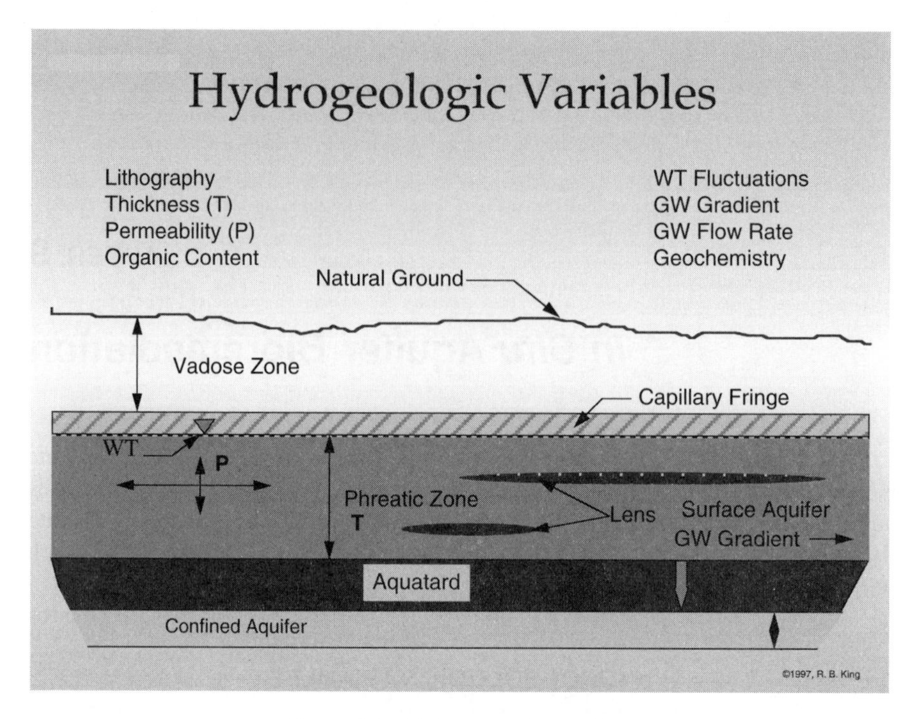

Figure 9-1 Hydrogeologic variables

LIGHT VS. DENSE NAPL

Nonaqueous phase liquids (NAPLs) come in two varieties for groundwater discussions. The light NAPL (LNAPL) variety consists of liquids that are less dense than water and do not enter the aqueous phase and so are not very soluble. Not being miscible and being lighter than water, they will float on the water table. We call them "floaters". On the other hand, a dense NAPL (DNAPL) is a liquid of low solubility that is more dense than water and will be found to sink below the water table. We call them "sinkers."

In the "classical" case, where soil and groundwater are contaminated with petroleum hydrocarbons (an LNAPL floater), soil contamination usually extends down to and at least 2 ft below the water table and may extend downgradient from the source for as much as a mile carried in the dissolved plume. A contaminated groundwater plume often extends even farther downgradient than the sorbed soil contaminants. If the hydrocarbon source is above the water table, the unsaturated zone will also contain quantities of hydrocarbons which will require cleanup to prevent recurring groundwater contamination.

Figure 9-2 shows a sketch of a typical LNAPL plume. It is important to note that the soil plume and the groundwater plume are distinctly separate in terms of extent, but that both must be addressed in the remediation design. The plume shown extending

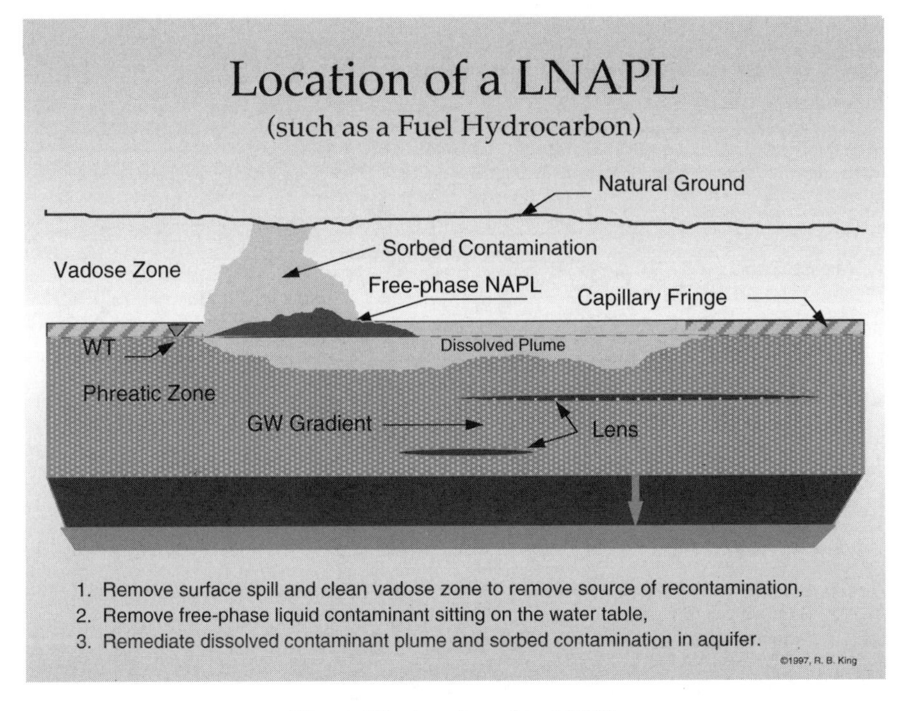

Figure 9-2 Location of an LNAPL

downward from the unsaturated zone can be the source of aquifer recontamination for many sites. If a separate floating liquid hydrocarbon phase (free-phase NAPL) is resting on the water table, this source must be addressed first in terms of removal.

When the offending contaminant is a DNAPL, it is common to find the aquifer polluted to a much greater depth due to the tendency of a DNAPL to sink in water. This situation will require remediation by cleaning deep aquifer soils and moving much deeper water than for a floating contaminant. Figure 9-3 gives a graphic illustration of the typical state of an aquifer that is contaminated by such DNAPLs as chlorinated solvents.

STANDARD PUMP-AND-TREAT OPTION

According to recent EPA and Oak Ridge National Laboratory statements, groundwater treatment by conventional pump-and-treat methods (used alone) will seldom lead to an effective clean closure.[1] Time in place for these systems often runs to 30 years or more without either effective treatment or closure. Even though surface treatment may be adequate for removal of hazardous constituents from the produced water, the final remediation of the aquifer being pumped is in doubt. *In situ* bioremediation is quite often retrofitted to these projects in order to speed the time to closure and ensure that aquifer soils have been fully remediated.

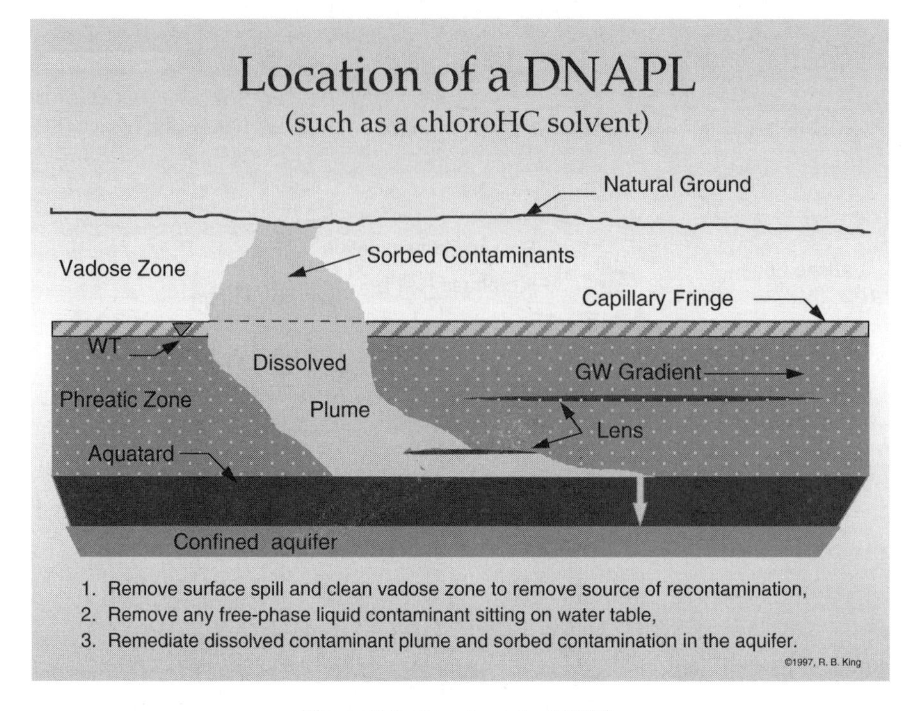

Figure 9-3 Location of a DNAPL

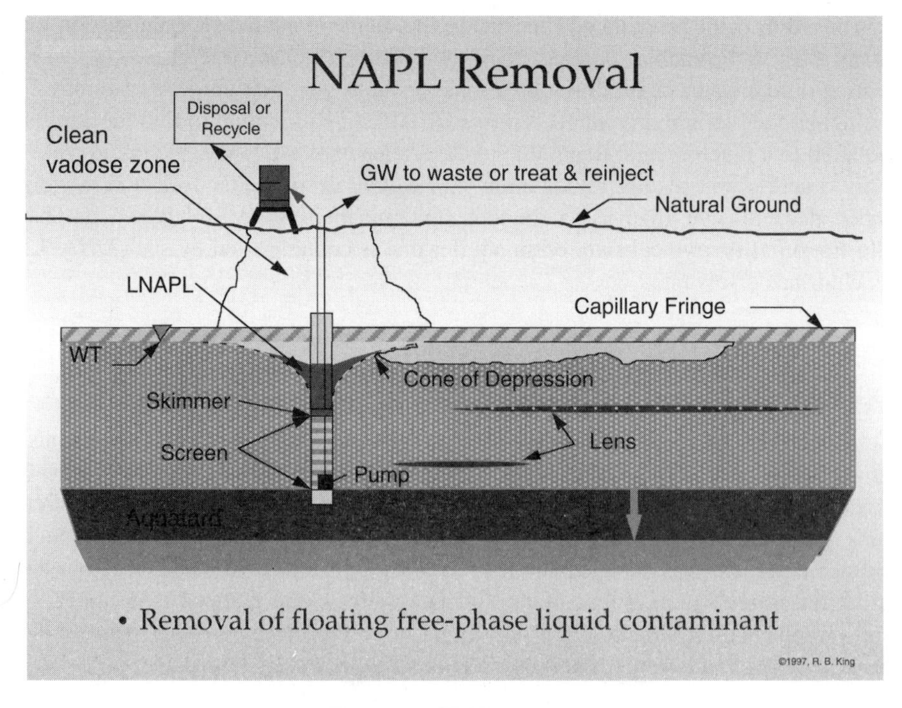

Figure 9-4 NAPL removal

IN SITU OPTION DESTROYS CONTAMINANTS

The benefit of *in situ* bioremediation of groundwater is that the contaminants can be destroyed in place. Both the groundwater and the aquifer soils undergo treatment simultaneously. Also, the destruction of pollutants carries with it the added advantage that liability is eliminated. Regulatory attention is turning toward those technologies that accomplish contaminant destruction, instead of mere removal to another medium or place.

Implementing a cleanup via *in situ* aquifer bioremediation requires a series of steps discussed here in the sequence found to be most successful by the authors. Taken in a typical sequence, they are:

1. Removal of free-phase NAPL
2. Removal/treatment of dissolved plume
3. Continued treatment of all affected aquifer soils

While other sequences are possible and some activities occur simultaneously, the logic will hopefully become apparent as we proceed.

NONAQUEOUS PHASE LIQUID (NAPL)

If an NAPL or free-phase hydrocarbon is present, the first activity should be to remove this liquid because it is the most concentrated source of hydrocarbon material. Hydrocarbon materials in high concentrations can pose a threat of recontamination. Previously treated media are sometimes toxic to microorganisms and are seldom biodegraded in free form.

The least expensive method for removal of free hydrocarbon is to pump out this liquid. Two-pump systems have been most commonly used, with one pump creating a cone of depression in the water table to act as a collection sump and the second pump or skimmer removing NAPL from the oil/water interface as shown in Figure 9-4. These pumps are typically installed in recovery wells near the center of the plume, since that is where the NAPL is usually concentrated. Several alternative methods have been developed and applied in recent years.[3]

The recovered liquid can sometimes be reused as a fuel or for product blending material, but often must be disposed of as a waste. The water pumped from the well almost always requires treatment (discussed elsewhere in this text) to remove the soluble organic materials prior to its disposal or reinjection.

HYDROGEOLOGY

Since we intend to use contaminated groundwater as our carrier fluid, we must first get the groundwater to flow at our command (see Figure 9-5). This requires the services of a competent hydrogeologist to properly design a system. The basic engineering design principles are covered here.

The initial site investigation will usually lead to the installation of sufficient groundwater monitoring wells to permit the local groundwater gradient and flow

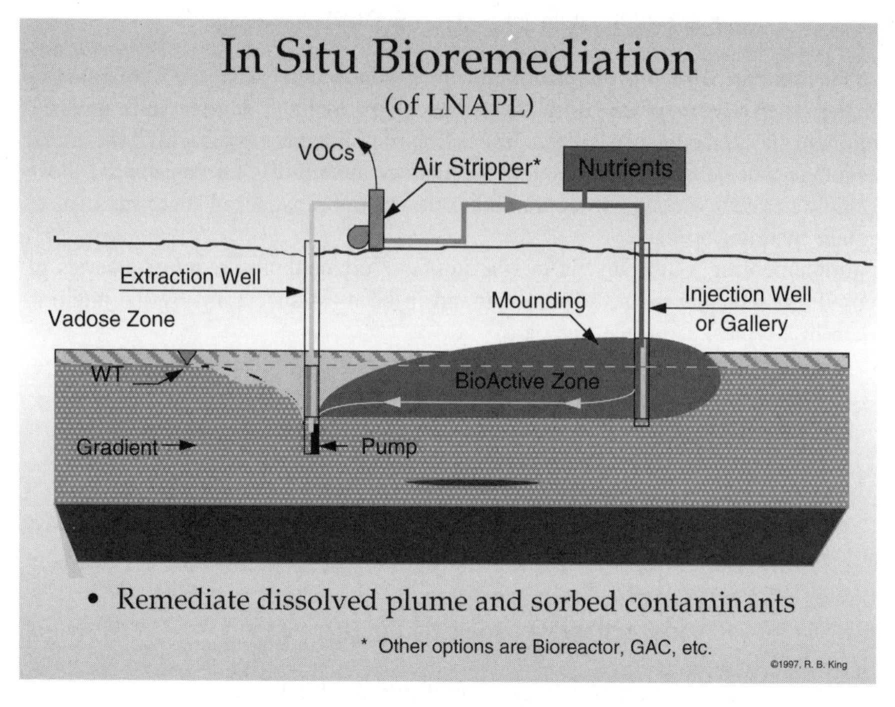

Figure 9-5 *In situ* bioremediation of LNAPL

direction to be approximately determined. The size of the contaminated groundwater plume will also need to be delineated. During the investigation and well installation, sufficient samples should be collected and analyzed to determine the approximate extent of the contaminated soil. This should include at least one soil sample below the water table for each well boring, and these samples will need to be specifically requested. Normal investigation practices in most states do not require that soil samples be collected below the water table. However, if these samples are not collected during the well installation, the extent of the soil contamination will have to be determined later (at considerable additional expense).

The installation of the wells must be performed under the supervision of a qualified geologist. While this is not a legal requirement in many states, the motive here is to produce a reliable log of the soil stratigraphy and character using a standardized method. The soil characteristics will have a profound effect on the design and operation of an *in situ* bioremediation system. The well logs produced during the installation are a valuable source of qualitative assessment information.

At this point, it is also useful to perform physical soil characterization tests to determine the values of parameters used in hydrogeological models. These include particle size, distribution, and porosity, and may include load capacity. In addition, any wells installed should be surveyed for location and elevation, using a reference benchmark on the site. The survey results are used for gridding purposes below and to provide an accurate picture of the groundwater gradient.

Some form of pump test will be required on at least one well installed in the contaminated zone. While this can be as simple as a slug test, where a dose of water is added abruptly to the well and the water level decline is measured as a function of time, a better method is a more formal pumping test. These tests range in duration from 6 hr to several months, depending on the soil type and the dynamic response of the soil/water system to a sustained pumping rate. The objective of the test is to determine the transmissivity of the formation as an average across the screened interval in the well. Transmissivity is a measure of the amount of water which can be passed through a given cross section of the subsurface.

The transmissivity and the groundwater gradient are used to calculate an approximate flow velocity across the site. It must be remembered that the calculated velocity represents average flow conditions across the depth of the well screen, and very little of the water may be flowing at the average rate if the soil is very inhomogeneous. However, the flow velocity provides an initial value from which to begin the design.

The normal design flow path between injection system and recovery system is taken as the long axis of the soil contaminant plume. Unless the soil is very coarse and the transmissivity is very high, the installation of pumping wells for water recovery will serve to retrieve most, if not all, of the contaminated water plume. This will need verification later when the groundwater modeling is performed, but, again, the design has to begin somewhere.

At this point, it is useful to determine the pore volume to be treated. This is defined as the water volume which exists between the injection and recovery systems. Initially the pore volume is calculated as the extent of the soil plume times the expected depth of recovery (the screened interval for the wells in the system) times the soil porosity. For "typical" contaminant loadings (up to several hundred ppm of hydrocarbon on the soil), in "typical" soil types (often sand or silty sand), the treatment will require the handling of 3 to 20 pore volumes of groundwater.

Given the groundwater flow velocity and the length of the soil plume, dividing the plume length by the water velocity gives a first approximation of the time required to process one pore volume of water. This is usually targeted to be between 1 and 3 months. A calculated time significantly longer than this would lead to consideration of multiple paths, to divide the treatment length into manageable sections. The trade-off between the cost of additional equipment is balanced against the shorter expected time required to complete the remediation. The preliminary time estimate for one pore volume is then multiplied by the expected number of pore volumes required (often a matter of "feel") to produce an initial estimate of the time required for the remediation.

This initial calculation should be used to give an impression of the scope of remediation required for the site. As long as the process still appears to be attractive, then hydrogeological modeling is normally undertaken. The preliminary data generated above are fed into a groundwater flow model in two or three dimensions (depending on the degree of sophistication required), and the effect of the injection and recovery methods on the flow rate and water gradient can be simulated. Trenches, wells, leach fields, and other methods to recover and reinject water can be modeled to determine the best methods to produce a rapid water velocity while containing the groundwater plume.

All of this hydrogeological effort thus produces four important results:

- An estimate of the required *pumping rate* to contain the plume
- An estimate of the capacity of the area to absorb reinjected water (*reinjection rate*)
- An approximate assessment of the *remediation time* required based on flow restrictions
- The *system geometry* required to produce the above values

MICROBIOLOGY

While the work described above is being conducted, a bioassessment study is usually performed to determine the behavior of the indigenous microbial population on soil and in water from the site under favorable conditions for biodegradation. The nature of the study is affected by the compromises necessary between exhaustive research and practical economics. Often the study is performed in two segments: an initial bioassessment or screening study to ascertain whether bioremediation is possible, which can be followed by a more detailed treatability study for determination of the most likely kinetics and degradation pathways for the materials found at the site, if it is required (see Chapter 4).

The initial assessment may take the following general form:

1. A composite soil sample is prepared from samples obtained from the contaminated zones. The average hydrocarbon content is measured, and specific compounds (volatile or semivolatile compounds) may be determined. The bacterial population is estimated from plate counts of total heterotrophic bacteria, and often a specific hydrocarbon degrading population is estimated. The degrader population is estimated using target contaminant vapors as the sole carbon source for laboratory cultures. An alternative method utilizes the most probable number (MPN) method. This method is more flexible and allows a better quantitative value to be obtained.
2. The same characterization described above is applied to groundwater from the contaminated zone. The water pH and other wastewater parameters such as suspended solids, biochemical and chemical oxygen demand (BOD and COD), total organic carbon (TOC), and background nutrient concentrations are measured.
3. Soil and water are combined in suitable flasks, usually at 10 to 25% solids, and treated with several ratios of nutrients and oxygen. A biotic poison is administered to provide at least one "killed" control sample to measure nonbiological effects of the treatment conditions (i.e., air stripping or chemical oxidation in the flasks). At periodic points during some predetermined treatment duration from 3 days to 2 weeks, the hydrocarbon content is measured to determine the approximate rate and degree of degradation. Other test parameters such as oxygen consumption, relative toxicity of the treatment conditions (measured by Microtox® or a similar apparatus), nutrient content, pH, and BOD may also be measured to provide an indication of the effect of the biodegradation under ideal conditions.
4. Geochemical testing is strongly recommended to determine the soil/water system response to the addition of nutrients and oxygen. These tests are necessary to avoid geochemical reactions which may cause site problems during the course of the remediation. One such problem is the precipitation of orthophosphate in high calcium soils or hard water. The objective of the test is to verify that the intended

injection concentration of phosphate (especially orthophosphate as the most suitable form for biodegradation) does not result in the precipitation of calcium phosphate that can clog well screens and perhaps the geological formation. Chelating agents may be added to the nutrient blend to minimize this problem, or complex phosphate forms may be used in higher concentrations to provide the essential nutrient.

5. Another geochemical effect has to be considered when hydrogen peroxide is being considered as the oxygen source. Iron, copper, and manganese will catalyze the decomposition of hydrogen peroxide into oxygen and water at a virtually instantaneous rate; this can pose a significant restriction on the injection concentration of peroxide which is possible without wasting the material and possibly causing outgassing in the formation. These metals can cause this effect at soil or water concentrations as low as 10 ppm. In the presence of these metals at or above these concentrations, it may be more practical to consider the use of oxygen to produce the optimum effect at minimum cost.
6. Where the degradation pathway requires detailed documentation, either to support a health-based alternative remediation target value or to satisfy fate and transport concerns of a regulatory agency, the frequency and type of analysis of the test flasks will be more involved and hence more expensive. These requirements may range from determination of the disappearance rate of specific compounds (i.e., polyaromatic hydrocarbons) using GC/MS techniques, to a determination of the full degradation pathway including the daughter products rising from the biodegradation of a suite of fuel constituents. Radiolabeling of target compounds can be an effective technique for studying contaminant breakdown sequences. In addition, the types of bacteria active in the degradation may be identified using staining or other suitable techniques, if it is a regulatory requirement.

It is important to remember that the test work described above is usually performed under ideal conditions (plenty of mixing, room temperature conditions, homogeneous soil and water conditions) and thus the apparent kinetics are likely to be quite a bit faster than those which are measured in the field. A "fudge factor" based on the experience of the particular group doing the work is applied to estimate the actual time required for field remediation based on the results of the treatability tests. This factor generally ranges from two to six, depending on the test conditions and the degree of confidence in the homogeneity of the site.

More detailed studies may duplicate the above work for a longer period to determine the degree of remediation obtainable (i.e., carry the degradation to completion in the laboratory), or one may conduct similar work at the incubation temperatures expected to occur in the field. Groundwater is typically between 50° and 60°F (10 to 15°C) in most parts of the U.S., so tests conducted in an incubator set to this temperature range can provide a more accurate picture of the process kinetics expected in the field.

ENGINEERING

When the foregoing work has been performed, a system must be designed to balance (1) the needs of the microbial process with (2) the ability of the aquifer formation to perform the transport function. The behavior of the nutrients in soil and water, the ability of the water to absorb oxygen from an appropriate source, and the need to manage the microbial population are balanced in setting the injection frequencies and concentrations of materials. The recovered water will also require some form of treatment at the surface in order to render it fit for reinjection. The materials of construction of operating components (pumping, injection, and treatment systems) need to be chosen carefully to ensure contaminant or waste matrix compatibility.

Perhaps the step in this sequence which is most likely to be neglected is the need to treat the recovered water. The successful start of microbial digestion of the contaminants usually produces an abrupt rise in the concentration of hydrocarbons in the recovered water. This is the natural result of contaminant desorption from aquifer soils due to the action of microbially produced surfactants. The degree of increase is dependent on site conditions and is not usually predictable. Thus, the hydrocarbon removal equipment at the surface needs to be designed to cope with a wide range of contaminant concentrations. As a starting point, the highest concentration observed in the water during the assessment phase may be used as the upper limit for hydrocarbons, unless some specific solubility factor is known and taken into account.

A variety of methods are available to remove hydrocarbons from water, including the biological processes described in Chapter 8. Some other processes which may be considered include air stripping, carbon adsorption, and chemical oxidation. The method chosen should produce an effluent quality that can satisfy the drinking water standard over the entire range of expected influent concentrations. This requirement that the method work essentially at drinking water standards represents starvation carbon levels for most biological systems. By this means, the microbes in residence in the soil/aquifer system are forced to utilize the target contaminants as food.

Neither air stripping nor carbon adsorption changes the chemical composition of the hydrocarbons. Stripping transfers them to the vapor phase and carbon attaches them to a solid. In either case, the material will still need to be chemically changed. Carbon is usually sent to an incinerator in batches, where the hydrocarbons are oxidized. The off gas from an air stripper may be treated using carbon (which is sent to an incinerator), or directly burned in a fume incinerator at the site. This requires additional processing and usually additional permits.

Whatever treatment process is chosen incurs additional capital and operating costs which must not be neglected in estimating the cleanup costs for the site. These costs can be a substantial fraction of the total cleanup cost. In practice, other alternative remediation technologies incur these costs over a longer period of time than that required for bioremediation, so that the process economics will usually tend to favor *in situ* treatment.

The design and operation of bioreactors for surface treatment of groundwater is covered in Chapter 8. Other commonly utilized supplemental cleanup systems are air strippers and carbon adsorption units.

Assuming that the cleanup system for the recycled water is set, the injection systems for nutrients and oxygen need to be designed. The nutrient blend is established as an output of the treatability study; this can either be a custom recipe or a standard mix of fertilizer components.[2] Depending on the amount of difficulty expected in keeping orthophosphate in solution, chelating agents (including complex phosphates) can be used in the recipe. The effects of the chelating agents should be tested on site water to demonstrate that the phosphate will not precipitate at the concentrations expected in the injected solution.[3]

New technology has made another oxygen supply alternative available. Formulated solid oxygen-release compounds are now available and in wide use in pellet and powder form for biological treatment of soils and ground water. This technology has the potential for reducing overall costs at many sites.[4] If hydrogen peroxide is used as the oxygen source, the solution strength planned for injection should be tested to determine that no outgassing occurs. As a practical matter, providing 35 to 40 mg/l of oxygen from either peroxide solution or bottled oxygen represents saturation at most field temperatures. The only practical way to exceed this concentration is the use of hydrogen peroxide under conditions where the decomposition rate is much slower than that created by catalytic decomposition.

The placement of all this material in the reinjection water will usually require an injection permit, and the conditions required for the permit will vary from state to state. Some states will require specific injection methods (i.e., use of injection galleries rather than wells) or limit the amount of material which can be injected to some specified quantity. Usually the largest constraint is that the water not attain a nitrate concentration exceeding 10 mg/l, which is a drinking water standard. The nutrient blend and quantities injected will need to take this limit into account.

Once the injection rates and frequencies have been determined, the equipment required to supply these materials can be designed. Here, three primary areas of concern surface:

1. Nutrient blends are usually prepared by dissolving the appropriate salts in groundwater from the recovery system. The degree of mixing and the solution recipe may sometimes lead to less than complete dissolution of the nutrient blend, and the choice of pump type can be important. Where possible, double diaphragm pumps will provide better service life, but have the disadvantage of pulsing the nutrients into the reinjection line and creating pulses in concentration through the feed system. While line snubbers can help to dampen the pulses, some vibration will be felt throughout the system, which adds mechanical stress.
2. Bottled oxygen and hydrogen peroxide are both strong oxidizers and require specialized materials of construction. Follow the supplier's recommendations! This is no place to attempt to save money; the proper containment and dispensing of the oxygen source is a major safety issue for these sites. Also, the provision of oxygen to the subsurface usually controls the degradation rate for the entire project.
3. These systems may have to operate below freezing in some climates, while the groundwater may still be warm enough to promote biodegradation while the surface is below freezing. Proper heat tracing will save many operating headaches, but the type used should be self-regulating to prevent temperature excursions above 110°F (43°C).

In addition, some consideration should be given to site security. This is, after all, a chemical operating system and should not be directly accessible to the public. Appropriate fencing, locks, and possibly guards may need to be considered if the site is normally operated in an unattended mode.

The above systems will require some form of power, usually electric. The use of a diaphragm pump will add the need for compressed air, which may also be used for the entire system if the risk of an explosion exists. This avoids the need for explosion-proof wiring and motors. Some systems may occasionally require cooling water. In general, the equipment specifications should include the utility requirements so that proper plans for their supply can be made.

INSTALLATION AND START-UP

The pumps, mixers, tanks, piping, sheds, and other equipment specified in the design are ordered and delivered to the site and assembled. The necessary utilities are brought to the site, the fences are installed, and the system is ready to start. A few hints on installation may help here:

1. Piping which is buried below the frost line does not need to be insulated or heat traced. It is also inaccessible to the public. The additional cost of subsurface piping may be regarded as a form of insurance against tampering.
2. Connections should be made with quick-connecting devices for convenience. As a remediation proceeds, sometimes the plumbing will need rearrangement, and these devices can save a lot of time and expense.
3. Check valves are very useful, especially in the oxygen source supply line. It is a critical safety issue not to allow contaminated water into a highly oxidizing environment; a good check valve can prevent this situation from occurring.

Usually the start-up sequence will follow the general outline indicated below:

1. The recovery pumps are started and the initial surge tanks are brought to the desired level. The water remediation system (bioreactor, air stripper, carbon beds, etc.) is started and brought to steady state. The effluent may be sent to some disposal site at first, until the treatment system is demonstrated to be working properly. When this has been done, the reinjection system is started to complete the water recirculation loop.
2. The first batch of nutrients is prepared and injected into the subsurface with the reinjection water. A conservative tracer may be added to the first batch of material to provide a means of measuring the groundwater flow rate directly; if the groundwater is low in chloride ion, salt may be added. If the chloride ion content is above approximately 50 mg/l, then sodium bromide may be added instead. It may be necessary to begin a project with low feed rates, which are gradually increased over time to monitor any precipitation or biofouling which may occur.
3. When the nutrient delivery system is working properly, then the oxygen source is started. This sequence depends heavily on the source material being used (usually the supplier of the oxidant will provide start-up instructions).

OPERATIONS AND MONITORING

After the system has been successfully started up, the operation enters the monitoring phase. While the details of monitoring each system will vary from site to site, the common elements to all remediations are given as follows:

1. The extraction wells across the remediation zone should be monitored in the field frequently for at least the first 3 weeks. During this time, any process problems associated with the equipment can be solved and the initial site readings can be taken. The field monitoring will include at least daily readings of temperature, pH, dissolved oxygen and nutrient content, chloride or bromide content, and oxidation/reduction potential (ORP) for selected wells across the site. These readings are used to calibrate the flow model used in the design to actual site conditions, and to indicate any problems in the behavior of the water pumping and injection systems.
2. Less frequently (anywhere from weekly to bimonthly), water samples from selected wells may be sent to a laboratory for selected analyses such as bacterial counts, MicrotoxR bioassay, oxygen uptake rate, or wastewater parameters.
3. After the first pore volume has been processed, the system should be carefully monitored to determine the changes brought on by the activation of bioremediation. The water treatment system may experience abrupt changes in hydrocarbon loading rate. Also, the introduction of nutrients and oxygen will normally produce a significant increase in the aqueous hydrocarbon levels and bacterial counts. Past experience suggests that a population of 10^6 colony-forming units (CFU) per milliliter of water is an optimum level; if the concentration gets appreciably higher, the well screens can become fouled with biomass at a relatively rapid rate.
4. As mentioned in Chapter 5, the onset of active microbial metabolism often causes increased hydrocarbon desorption and solubilization due to the production of microbial surfactants in the aquifer. The significance of this should not be lost as it constitutes direct evidence that the sorbed organic loading is being desorbed and made available for surface removal and microbial degradation precisely as predicted. **UNDER NO CIRCUMSTANCE** is this sudden spike in observed contaminant levels to be interpreted as a worse condition than existed prior to treatment or any other such nonsense. In fact, the hydrocarbon spike is your assurance that the remediation has started off with a BANG and should continue at an accelerated pace all the way to closure. As not all microbial communities produce aggressive surfactants, not all *in situ* aquifer remediations will experience this phenomenon. Indeed, some projects run to completion without any spike at all and show a steady but sure degradation. Just remain aware that it can happen and understand what it means. You may want to ensure that all parties are informed beforehand to avoid the obvious potential for heartburn if it comes as a surprise.

Experience also suggests that the bacterial population is limited by the supply of nutrients, and that the oxygen level should be maintained as high as practical until the recovery wells have reached the oxygen saturation value. Thus the design range of bacterial population counts is controlled by regulating the nutrient injection rate. A little experience with each system will guide the operations personnel in making proper nutrient additions to maintain the desired bacterial population levels.

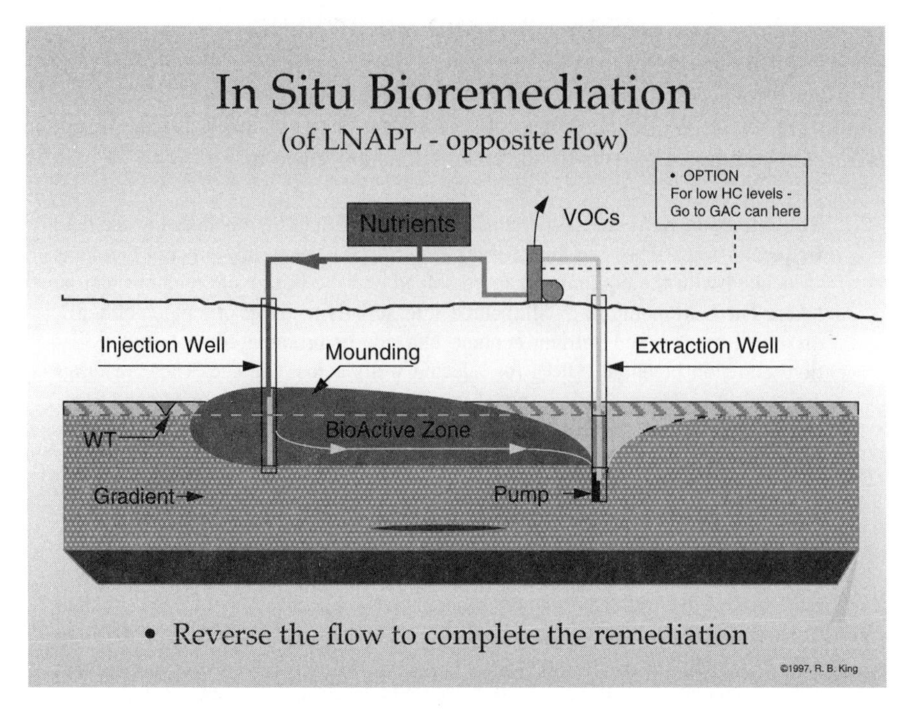

Figure 9-6 *In situ* bioremediation of LNAPL — opposite flow

Some mechanism for removing biomass from the injection and recovery wells is still likely to be needed over the course of the remediation. If hydrogen peroxide is the oxygen source, a 0.5% peroxide solution poured into the well and allowed to sit for 4 hr has been found to be effective. At the end of this time, the well is pumped and the process is restarted. If another oxygen supply is used, bleach may be used to clean the well in the same fashion.

When the operating parameter monitoring for an LNAPL remediation indicates that initial cleaning has taken place, the injection and extraction strategy can be reversed in order to clean the remainder of the aquifer soils (see Figure 9-6). Where the original injection well soils have been cleaned through mounding of the injection stream, these bioactive fluids will have removed or destroyed the contamination within their flow path. Reversing the flow will now place the bioactive fluids into contact with the contaminated soils that remain close to the original extraction well.

Where *in situ* aquifer bioremediation has been applied for remediation of a DNAPL, the initial extraction of the dissolved plume can usually follow a simpler flow path and sequence of events. The initial plume removal (Figure 9-7) can be followed by injection of bioactive fluids that follow the original route of contamination. These bioactive fluids can be induced to follow the original flow path of the contaminants through careful engineering of extraction/injection locations and flow rates. Figure 9-8 illustrates the screened interval and locations for the desired movement of fluids.

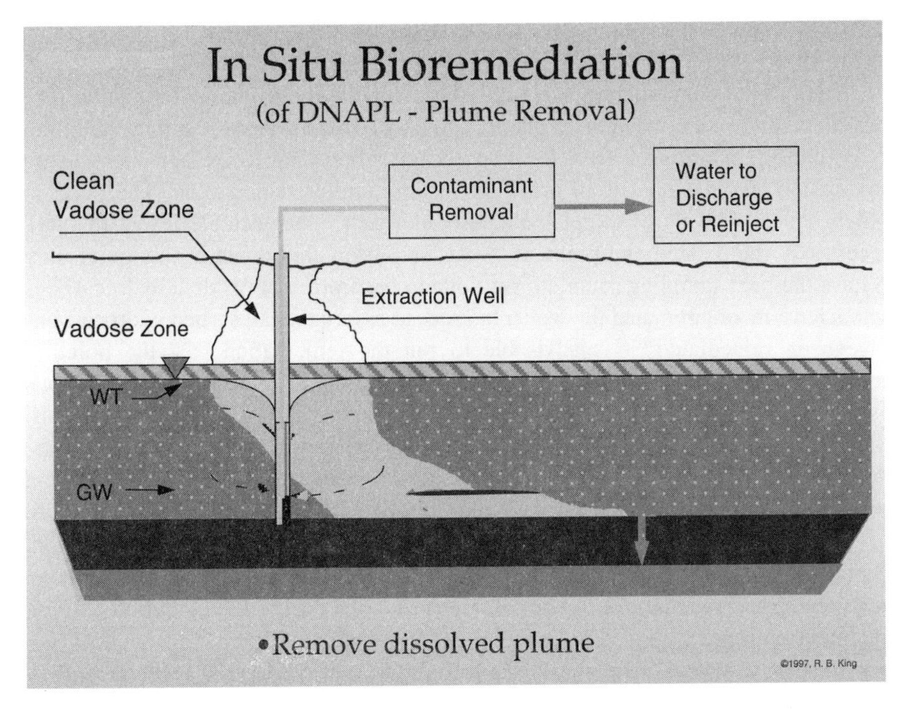

Figure 9-7 *In situ* bioremediation of DNAPL — plume removal

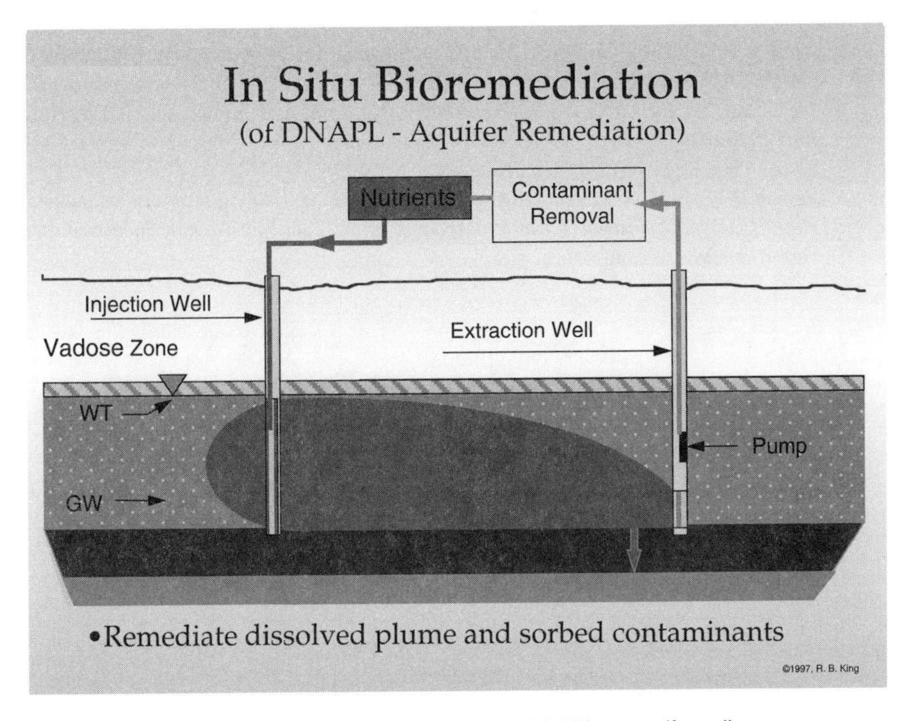

Figure 9-8 *In situ* bioremediation of DNAPL — aquifer soils

CLOSURE

When target contaminant levels fall below regulatory levels or some other pre-negotiated end point, the closure plan is activated. This amounts to a final sampling round and preparation of necessary reports for submission to the agencies having jurisdiction. A final note is in order for *in situ* aquifer remediation. The point at which aqueous levels of target hydrocarbons reach nondetectable levels IS NOT necessarily the desired end point. Projects have been shut down at this point only to have the next sampling round go sour due to desorption of residual hydrocarbons which have recontaminated the wells. In order to assure that all sorbed hydrocarbons have been remediated, it is advisable to run the remediation past the point of nondetect on the targets and continue the operation until the oxygen and nutrients rise to injection-equivalent levels and microbial counts fall to pretreatment levels. At this point, all sorbed hydrocarbons have been removed and degraded (and the subsurface is depleted of microbial food) and the wells can be shut in. Subsequent postoperational sampling should then yield consistent negative results for organics.

These steps serve as a general guide to the practice of *in situ* bioremediation. There are two special cases that remain: vadose zone bioremediation and lagoon treatment. Special situations can be encountered in the design and operation of these treatment options.

REFERENCES

1. **Anon.,** 1990, Pump-and-treat method ineffective for contaminated groundwater, *HMCRI Focus*, November, p. 3.
2. McMillan, D., Desert Green Bioremediation Nutrients and Equipment, Belen, NM, (505)864-2829.
3. Boyd, T.R., Aqua Process, Inc., Houston, TX, (409)265-1710.
4. Fulton, D.E., 1996, Selection and application of effective LNAPL recovery techniques. *Hazardous and Industrial Waste.* Proceedings of the 28th Mid-Atlantic Industrial and Hazardous Waste Conference, p. 619.

CHAPTER 10

Lagoon Bioremediation

Bioremediation technology can be tailored for effective treatment where an appreciable amount of water cover exists in a contaminated lagoon. Generally, the lagoon also contains a sludge (or other fine solids), which is the primary source of hydrocarbons or other organic contaminants.

TREATMENT OPTIONS

Those contaminants that have entered the dissolved phase in the lagoon water can be easily treated through aeration and mixing. Oxygen for aeration comes from air which is blown into the water column in the lagoon using commercial wastewater aeration equipment. These industrial aerators can have an effective surface aeration radius in excess of 50 ft. When a separate aeration lagoon is so constructed for wastewater treatment by aeration, the industrial process (and the unit) is sometimes referred to as BIOX (for *BI*ological *OX*idation).

Lagoons containing hydrocarbon-impacted sludge are often found in petroleum refineries, paper mills, chemical plants, and as old disposal pits. The nature of the hydrocarbons differs according to the original use of the lagoon. Since many of the lagoons were also used to process the storm runoff water from the facility which can generate high quantities of contamination, the concentration of hydrocarbons can vary from hundreds of milligrams/kilogram (ppm) in the sludge to over 20% by weight. A "typical" lagoon profile is illustrated in Figure 10-1. Most of the lagoons described above fit this profile fairly well.

Although the water phase can be easily treated via aeration, its recontamination by the sludge and lagoon solids constitutes the major focus for lagoon remediation. There are two basic approaches that can be used to treat the lagoon solids or sludge:

1. Treat and remove the water, and then use *land treatment* methods on the solids. This is most effective for low hydrocarbon concentrations and situations where the material is not classified as hazardous by RCRA standards. Land treatment is addressed in Chapter 7.

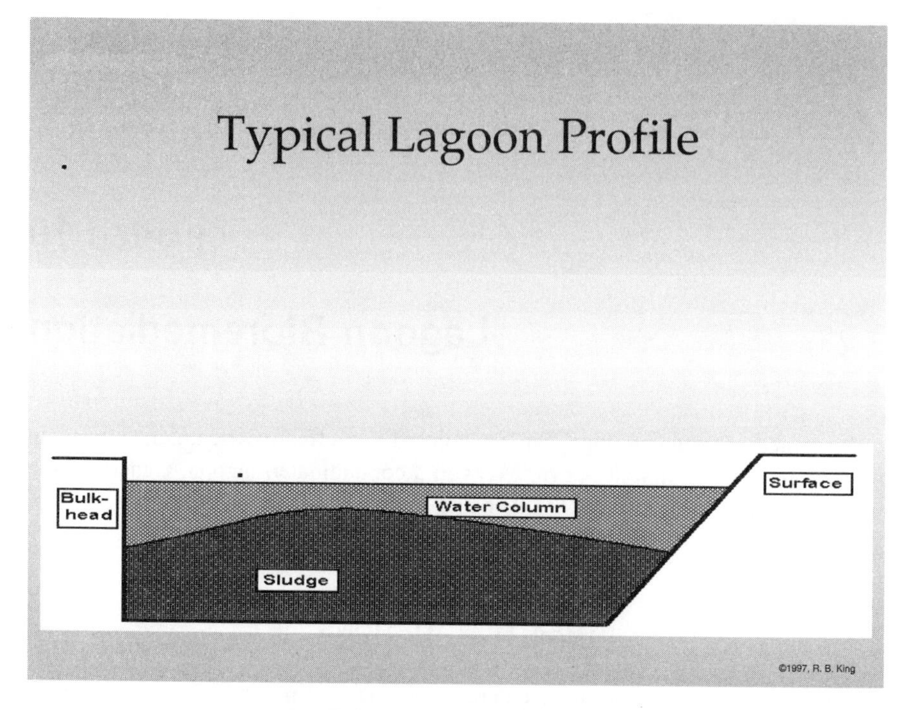

Figure 10-1 Typical Lagoon Profile

2. Treat the entire lagoon as an *in situ bioreactor* by moving the highly contaminated sludge or solids into the aerated water column. This constitutes an *in situ* method for some hazardous materials and may allow an effective strategy for management of the toxicity issues raised by very high hydrocarbon concentrations.

This method of lagoon bioremediation is more capital (equipment) intensive than simple land treatment, and thus is more expensive on a unit basis. It is not likely to be cost effective unless the material is RCRA hazardous, other disposal methods are unavailable, or the customer's policy requires destructive methods of cleanup to eliminate future liability.

FEASIBILITY TESTING

Given the possibility of wide variations in the type and concentration of hydrocarbons and organics requiring treatment, *feasibility testing* should be considered mandatory. This can initially take the form of low dosage *shake flask tests* using 5 to 10% sludge in lagoon water, and suitable nutrient amendments to achieve a carbon to nitrogen to phosphorus *(C:N:P) ratio* of approximately 400:10:1. Concentration profiling and hydrocarbon characterization will normally be done during the delineation phase of the project; remember that the sludge lagoon may have impacted its immediate surroundings, and don't forget that you may have to remediate the surrounding soil and groundwater, too. These areas may be treated using the other methods described in this book.

The shake flasks are normally monitored for microbial populations of total heterotrophic bacteria and hydrocarbon degraders, toxicity using a method such as Microtox™ to determine whether the hydrocarbon dosage is too high, nutrient uptake from solution (as measured by remaining nutrient soluble concentrations), and oxygen consumption or CO_2 production (either using respirometry or oxygen uptake rate [OUR] methods typically used for wastewater). Contaminant concentrations may also be tracked to obtain an approximate degradation kinetic.

An extended bioreactor study also needs to be conducted to determine the growth and acclimation of the indigenous microbial consortium which can likely be achieved once the full system is operating. This is done by increasing the sludge loading in a continuous bioreactor in stages and measuring the point at which hydrocarbon toxicity begins to limit the degradation rate. In some well-acclimated systems, the hydrocarbons may be found to be degraded effectively at any suspended sludge concentration. However, most systems are likely to be limited by the toxicity of the suspended sludge. This case is assumed in the description below, as toxicity management is critical to the success of the remediation.

DESIGN CONSIDERATIONS

Bioreactor tests as described above should provide the following design information:

1. The oxygen consumption rate, based on the respirometry or OUR results
2. The nutrient loading rates expected for the full system
3. The sludge loading rate which can be sustained in the lagoon

The sludge loading rate is critical for toxicity control. The lagoon is operated as a bioreactor of increasing volume, and the sludge (being loaded into the water column from below) must be added at a controlled rate so that the toxicity is controlled.

Lagoon bioremediation requires mechanical systems for oxygen injection, sludge suspension, sludge shearing, and nutrient supply. Each will be described separately, although system design must provide for their coordinated interaction for the remediation to proceed.

OXYGEN INJECTION

Oxygen dissolution is usually the rate-limiting factor in the process design. Oxygen must be dissolved in the water to be usable for bioremediation. Aeration produces dissolved oxygen in water at only about 8 mg/l at saturation. This rather low oxygen availability will provide a small degradation kinetic. Therefore, the oxygen concentration (a factor of aeration rate) should be held as high as possible without creating the problem of air emissions or excessive horsepower consumption costs.

The respirometry or OUR value is usually expressed in milligrams/liter/minute of oxygen consumption, and values of 5 to 20 are typical for lagoons treating hazardous wastes. The oxygen dissolution rate should match the OUR from the bioreactor test at its maximum value, so that O_2 dissolution does not become the rate-limiting step. The air movement rate is NOT the oxygen dissolution rate — an aerator with an efficiency of 5% oxygen dissolution at an OUR of 10 mg/l/min requires: (10 mg O_2 dissolved/liter of water/minute)/(0.21 mg O_2 injected/milligram of air)/(0.05 mg O_2 dissolved/milligram O_2 injected) = 952 mg air per minute per liter of water. In more conventional process units, this is 0.10 cfm air per gallon of water in the lagoon. For a one-acre lagoon with 4 ft of water cover, this is 130,000 cfm of air.

Air injection equipment is readily available in a wide variety of designs, based on wastewater treatment needs. This equipment generally forces air bubbles into the water column and is limited by the contact time maintained between the air bubbles and the water column for dissolution of atmospheric oxygen into the water. The dissolution efficiency, especially for shallow systems, is relatively poor, so some of these methods can only dissolve 5% or so of the oxygen injected into the water. If the oxygen is not dissolved in the water, it is unavailable for degradation and is thus wasted. The price of air is free. However, the horsepower is not, and these electric motor-driven units tend to consume a lot of power. In addition, they generate large amounts of froth from the air which is not dissolved, and can strip volatile organic compounds from the water column into the atmosphere. Downdraft aerators have been used in this service with relatively low VOC emissions by placing the units as deep as possible in the water column.

The major limitation of this method is that air can dissolve oxygen at a maximum of approximately 8 mg/l in water. This effectively limits the degradation rate, since oxygen is usually the limiting constituent. These limitations can be somewhat overcome by injecting pure oxygen. Pure oxygen has a solubility of approximately 20 mg/l in water because there is no nitrogen from air to compete for interstitial space. This commodity chemical is readily available (in liquid form for large quantity consumers) may be generated on site (often at less cost), and may be completely dissolved in systems designed for this service. The oxygen is injected into a side-stream from the lagoon and permitted to travel in a closed system (pipe) for a sufficient length of time to dissolve most of it. The pumped water is then injected below the surface of the lagoon through a venturi nozzle to complete the dissolution and also mix the water column. This method permits the system to operate at dissolved oxygen concentrations up to 15 mg/l, which effectively doubles the delivery (and degradation) rate. Its use in large lagoon bioremediations typically has a total unit cost of approximately $1/cubic yard of sludge treated, even while starting at over 20% hydrocarbon in the sludge. For smaller systems the unit cost would be higher because the system capital cost is relatively large.

SLUDGE SUSPENSION

The second critical step is the proper suspension of the sludge in the water column. This loading is best performed by gentle lifting and mixing. Sludge lifting is performed using low-speed mixers with large blade areas. "Banana blade" mixers have been used successfully in this service, turning at 2 to 3 rpm. These mixers lift the sludge from the top of the sludge layer and push it up into the water column, where the aerator system helps to keep it suspended by mixing the water in the lagoon.

The sludge density has a direct bearing on the ability of the sludge to stay in suspension and, hence, the amount of low rpm mixing which is required. Dense sludge particles will settle out of the water column faster, and require more mixing energy to keep the optimum amount of sludge in the water column (the bioreactor). This density should be measured as the true particle density, not the bulk density of the sludge. The shearing action described below serves to break the sludge into individual particles, and the sludge bulk density is usually measured with a high water content. Drying the sludge prior to measuring the density is important for this design element.

The mixers are typically suspended from flotation devices which are capable of being moved around the lagoon. They must be capable of digging sludge out of corners, around obstructions, and ideally out of any large pipes which enter the lagoon. The flotation devices range from pontoon systems to dredge barges, depending on the size of the lagoon.

The mixers must be capable of lifting sludge from the full depth of the sludge column since, as the remediation proceeds, the sludge to be lifted will be deeper and deeper below the water surface. If the soil below the lagoon is to be treated using the same system, this must also be taken into account when sizing the blades and flotation devices.

SLUDGE SHEARING

The last design element is the shearing of the sludge. This is important to create a large surface area for effective microbial colonization of the solids. Unlike many bioremediation methods, this form of treatment relies on direct colonization of the solids upon which the contaminants are adsorbed. The sludge must be separated into very fine particles for microbial colonization to be effective. This function is ideally performed using different equipment from that used in lifting the sludge.

Typically, the shearing is performed with an open-face impeller centrifugal pump, usually with a screen mounted over the face to prevent large items from blocking the flow. The pump is usually mounted on a pontoon device, although very large systems may use a barge. The pontoon is then equipped with a motor to raise and lower the pump into the water column. As with the sludge suspension, the pump should be able to shear sludge near the bottom of the sludge column, so it is usually mounted on the bottom end of a riser pipe of sufficient length to reach through the water column to the base of the sludge layer. The pump is gradually lowered into the water as the sludge is lifted and operates ideally just above the surface of the sludge layer.

When relatively small lagoons are being treated, the shear pump may be used to lift and suspend the sludge as well as to shear it. While this can be done, it slows the remediation in practice, because the pump location for lifting sludge is different from the best location for shearing.

NUTRIENT SUPPLY

Solid nutrients in the form of commercial fertilizers may be broadcast across the lagoon or dissolved in lagoon water in a small independent mix tank and pumping system. They may be injected into the water recirculation system when pure oxygen is being injected by that means. If orthophosphate is used (the preferred microbial form of phosphorus), any calcium in the water will precipitate it fairly rapidly. Other phosphate forms are not as bioavailable but are easier to keep in solution. The preferred form of nitrogen is ammonium ion, as long as the pH is kept under control. Ammonium salts are available in commercial fertilizers and some nitrate may be present. Careful selection of nutrients for solubility and potency is essential and may vary considerably from site to site.[1]

OPERATION AND MONITORING

Starting the system operation requires the early development of sufficient microbial biomass in the lagoon to perform the remediation. In order to obtain these conditions, the oxygen and nutrient systems are started first, and then the low speed mixers dose an initial load of sludge into the water column. The shear pumps are started and the water column monitoring begins. Samples are usually tested onsite for OUR, dissolved oxygen, ammonium nitrogen and orthophosphate, and pH using test kits and portable meters. The water temperature is also monitored for correlation with OUR. Samples may also be sent to a laboratory on a periodic basis (more frequently at start-up) and tested for total suspended solids (TSS), total dissolved solids (TDS), and volatile suspended solids (VSS) as a crude measure of the biomass vs. sludge concentration, and some measure of hydrocarbon content — often total organic carbon (TOC). Care must be taken in interpreting the carbon results because the suspended matter normally contains most of the carbon (as adsorbed contaminants), and thus the result is impacted by the TSS content.

Toxicity testing is also suggested as a means to judge when to add more sludge to the water column. These systems can be started using Microtox™ as a process control parameter. Plate counts for total heterotrophic bacteria and hydrocarbon degraders may also be monitored as an indication of the health of the microbial population.

Once a biomass has been established, the process consists of operating the sludge suspension mixers to dose the sludge into the water column as fast as the biomass can acclimate and degrade it. This means that the remediation may start slowly, but it should accelerate all through the process as long as the temperature remains high enough and the toxicity is properly managed. The shear pump operates at a relatively constant speed and the oxygen delivery system injects oxygen to maintain a preset *dissolved oxygen* (DO) value. Toward the end of the remediation, this value may drop somewhat as the system delivery rate reaches a maximum, but a significant drop during the remediation may indicate a design problem or an equipment problem with an aerator.

As indicated above, air monitoring will probably be required in order to obtain an operating permit from the regulatory agency. Both types of oxygen delivery systems described have been successfully operated within their air permit limits, but the pure oxygen injection system commonly has no measurable concentration of volatile hydrocarbons released to the atmosphere. The reduction in volatile emissions was estimated to be greater than 99% using this method.

CLOSURE

For many lagoons in present or former refinery service, residual metals in the sediments remaining after bioremediation will need to be addressed. These may be treated using fixation or stabilization techniques, or covered and capped in place as part of the closure plan. Even if offsite treatment is required for residual metals, the reduction in sludge volume and toxicity effected by bioremediation will usually save considerably more than the costs for operation. RCRA hazardous sludges may complete the treatment as nonhazardous materials, which opens other disposal options (often at much lower costs).

In all these techniques, the final test of the success of the remediation is residual sludge and soil sampling. Ultimately, the site soil/sludge is the source of the contaminants, and most agencies will require a demonstration that the site soil has been cleaned prior to release from further work.

REFERENCES

1. McMillan, D., Desert Green Bioremediation Nutrients and Equipment, Belen, NM, (505)864-2829.

CHAPTER 11

Vadose Zone Bioremediation

Whenever soils above the water table (the vadose zone) are contaminated with biodegradable organic pollutant materials, an opportunity may exist for application of one of the proven methods for vadose zone bioremediation. The several methods available generally provide sufficient oxygen and nutrients for native or augmented microbes to carry out the natural degradation of the pollutants. These methods have received considerable attention of late, due to the rising costs associated with environmental cleanup by site owners as well as the government-funded programs such as SuperFund. Bioremediation, as a site restoration method, is broadly viewed as one of the least expensive means of achieving environmental compliance in recent years. Recent advances in vadose zone applications mandate a separate chapter to address the intricacies of this technology.

A few years ago, these methods were considered to be highly innovative and suspect. Over the past 15 years, the methods described herein have become not only commonplace, but preferred by many professionals and regulatory agencies.

BIOVENTING

In cases where hydrocarbon contamination is found above the water table, while some soil moisture is always present, the affected area will contain soil pore spaces also filled at least partially with air. This open-pore volume may be used as the carrier medium for air movement for the delivery of oxygen, and for the removal of volatile contaminants which reside in the pore spaces as soil vapor. This process thus bears a resemblance to *soil vapor extraction* (SVE), in which air is used as the carrier to remove volatile contaminants from the subsurface. However, since the purpose of the air movement in bioventing is to deliver oxygen rather than to remove volatile compounds, the flow rate required is often substantially lower than would be used in an SVE design. In fact, a primary objective of bioventing is to degrade as much of the hydrocarbon as possible in-place in the subsurface, so it does not have to be treated at the surface.

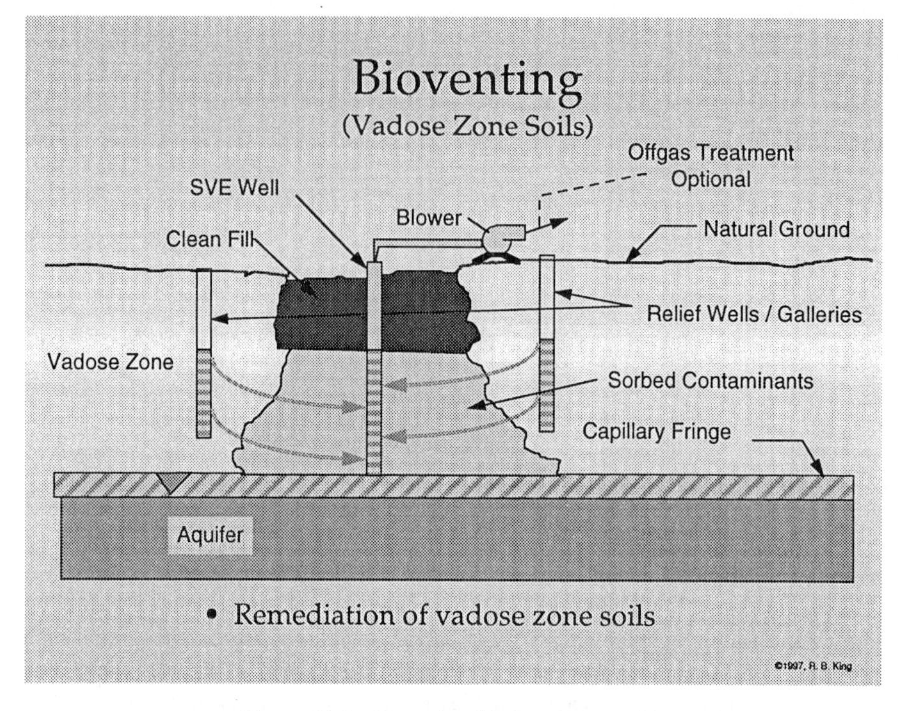

Figure 11-1 Bioventing (Vadose Zone Soils)

The basic microbial principles are the same as for the other forms of bioremediation. We need to optimize the local subsurface conditions to the extent possible in order to maximize the growth of aerobic hydrocarbon-degrading microbes. This is done by designing and installing air and nutrient/water management systems to deliver the necessary materials to the impacted zone and control the removal rate of compounds out of the local geologic system. The objective is to reach an agreed treatment standard for the soil as expeditiously as possible, and then stop.

The air brought to the surface may require some treatment to remove any residual contaminants before it is released to the atmosphere (or reinjected into the ground without treatment). This treatment can be either carbon adsorption of the vapor or fume incineration (often using a catalyst to lower the required combustion temperature). Vapor-phase bioreactors (biofilters) are being employed in this service with increasing frequency, as the method becomes more accepted. In Europe, these biofilters have been available for almost 20 years.

A "generic" bioventing setup for a vadose zone hydrocarbon plume is shown in Figure 11-1. The illustration does not show the case where the groundwater below the vadose zone plume has been impacted. Bioremediation technique for removal of contamination below the water table requires a completely different engineering design. Aquifer bioremediation is addressed in Chapter 9.

Delineation

Delineation activities for vadose zone contamination are almost exclusively directed by data from soil samples. Soil gas surveys may be performed to provide an initial set of locations for soil sampling, but the contaminant type, quantity calculations, and concentration data require soil samples for proper accuracy. It is important to obtain data from various depths all the way down to the water table in order to characterize the extent of the impacted area and the mass of hydrocarbon to be treated, as well as to determine whether the site groundwater has been affected. However, in areas where the water table is very deep and the impacted zone is relatively shallow, there is usually no need for groundwater monitoring wells in the delineation phase.

Soil texture is a critical parameter that needs measurement during delineation activities and is as important as knowing the concentration and extent of contamination, because it will indicate how readily air can be moved through the subsurface. In general, coarse-grained sands and gravels permit air to move more readily than the fine-grained silts and clays. Generally, it is very difficult to move air (or water) through these. However, air is considerably easier to transport than water, and silty soils will often permit adequate air movement for bioventing. As with other forms of bioremediation, nutrient and microbial characterization information (described in Chapter 4) is essential in order to gauge how well bioremediation is likely to work.

Once the plume shape, volume of soil, and mass of the pollutant contaminants to be treated have been determined, the equipment and engineering design phase begins. Additional design information will be required in order to properly size the equipment. This generally takes the form of a respirometry test.

Respirometry

This testing provides a general indication of the rate of microbial oxygen consumption or CO^2 generation in a closed microcosm tube under simulated remediation conditions. This knowledge indicates the amount of air that needs to be moved during remediation so that the blower may be sized correctly. In the past, many bioventing systems were designed on the basis of "rules of thumb" used for vapor extraction systems which must move considerably more air than during bioventing. As a result, much more installed horsepower and air movement were used than necessary to support the remediation through bioventing. This misapplication of the technique often required more surface treatment than necessary, expensive equipment, and higher flow rates. A respirometry test will facilitate sizing of the air flow requirement, save considerable installed cost, and still accomplish the remediation in the same time scale.

One field treatability test procedure developed for the Air Force[1] consists of two principal parts:

1. A flow characterization test using a tracer gas (such as helium) to measure flow and dilution rates at various points in the subsurface flow pattern.

2. A respirometry test which consists of air injection with a dilute tracer gas (often helium for flow rate detection) for a specified period (perhaps 24 hr), followed by oxygen and carbon dioxide measurements of the soil gas at various points under no-flow conditions. The rate of oxygen disappearance is converted to a degradation rate. All the data reduction must be based on assumptions for contaminant structure (hydrocarbons, in this case) and degradation efficiency. The field data usually contain scatter, but an average value is used for the sitewide design.

While this approach has its drawbacks in terms of representing the entire site and all the subsurface strata, it has more validity than simply moving all the soil air that is possible with a vacuum pump or blower and flooding the soil with atmospheric oxygen. In addition, the duration of the treatment may be estimated by dividing the calculated degradation rate in mass per unit time into the hydrocarbon mass present in the soil. CAUTION: This is only a first estimate and does not take into account the reduction in degradation rate that occurs whenever the hydrocarbon concentration falls to starvation levels for the microbial consortium at work. It also does not account for increases in rate for "hot spots" where the contaminant concentration provides an increase in the degradation rate. It should, however, provide an estimate of the time (e.g., weeks, months, or years) likely to be involved, and can assist the engineer in choosing the capacities of equipment that will be required.

The protocol described above usually provides an estimate of the *radius of influence* for the vacuum induced. Since this term is adapted primarily from vapor extraction tests, it is principally applied to vacuum pump tests. The radius of influence is taken as the point at which a given vacuum drawn on the extraction well produces a measurable effect (generally 1 in. or so of water gauge vacuum) on a *piezometer* some distance away from the well. The test is usually conducted with several piezometers at varying distances from the extraction well, and a graph of induced vacuum vs. distance from the suction well provides the radius of influence. This graph is likely to have a considerable amount of data scatter as well, but the trend can be statistically normalized.

A few words of caution are in order: the induction of a vacuum provides no information as to the actual flow rate through the soil. Even 1 in. of water gauge vacuum is optimistic in terms of significant air movement at the outer perimeter of the apparent radius of influence. Since the mass flow of air (volume) is the same at the outer perimeter of the radius of influence as it is in the extraction well, as the swept soil volume increases (with distance from the well), the air flow velocity decreases as the square of the distance; the air flow velocity at the outer radius can be essentially zero.

Design Considerations

The system design usually requires an iterative process to establish the design air flow rate. A critical decision must be reached at the beginning of the process: do we want to *inject or extract* the air? Each has its own advantages and disadvantages. Injection eliminates the need for surface air treatment, but the air must be contained in the subsurface until the hydrocarbon content is degraded, which may require the flow rate to be unacceptably slow in shallow systems. Injection also carries an additional liability; contamination can be spread to unaffected areas and adjacent properties if the flow is not completely controlled. Air extraction, on the other hand, permits better flow rate freedom and higher assurance that the contaminants that reach the surface are contained for treatment. This also requires the additional cost of a treatment system for air at the surface. Different branches of the armed forces, as well as individual designers, have different preferences and may prefer air injection to maximize the subsurface degradation. There is no strict right or wrong way. The decision should be partly based on the ability to obtain an operating permit from the appropriate agency and to keep the discharge of contaminants to the ambient air below regulated concentrations, while providing enough oxygen to promote aerobic biodegradation.

The well spacing is usually taken as equal to the radius of influence as a starting point. The area to be treated can be covered with wells having enough overlap to move air through any apparent blind spots. Some designers will use the pilot test data to calculate an air permeability and insert the value into a vapor extraction model (analogous to groundwater modeling) to simulate air flow through the soil. This approach has a sound technical basis to determine the subsurface layout, but it is more expensive than the traditional "radius" approach. There are several appropriate models for vapor extraction (Hyperventilate™ is one of the more popular), and these models are being adapted for prediction of bioventing behavior. They can also predict the extracted air contaminant concentration for vacuum systems, or the resulting ambient air concentrations above an injection system.

A model developed for the U.S. Army Corps of Engineers uses both injection and extraction across the same treatment zone, and this produces a significant reduction in the required flow rate. For the model results to be valid, the *injected flow* into the ground must balance the *extraction flow*.

The flow rate is established either from the modeling results or from the respirometry test on a per-well basis, and this value is multiplied times the number of wells to obtain the *design air flow rate*. This may be an iterative process as the number of wells and their layout are adjusted to obtain the most even flow coverage with the minimum number of wells.

Once the flow direction and rate have been selected, the appropriate equipment are chosen to deliver (or remove) the air. Either low pressure air compressors, conventional plant air compressors with storage tanks and pressure regulators, or vacuum pumps (usually regenerative blowers or rotary lobe pumps) are sized to move the required amount of air. The extraction or injection wells are placed as close as possible to the center of the soil plume and spaced according to the test results described above.

For air injection, conventional plant air supply compressors are robust, easily obtained, and reasonably inexpensive. The incoming air is usually filtered using a pleated-paper cartridge unit. An air regulator is required to control the injection pressure. These compressors can either inject air at low flow rates continuously, or can "spike" inject into the subsurface on an intermittent schedule (e.g., 8 hr per day). As the biologically available carbon decreases as a result of bioremediation, the rate or frequency of injection can be reduced accordingly as the oxygen demand for degradation is reduced.

For air extraction in many soils regenerative blowers are appropriate, readily available, and relatively inexpensive. They require inlet filters, often a cyclonic separator to remove water droplets from the incoming stream, a pleated filter, and a silencer on the discharge to keep the equipment noise at acceptable levels. Since this extracted air stream often requires treatment, the pump may be sized with sufficient pressure capacity to deliver the extracted air through the treatment unit.

For shallow systems, the air piping may be installed in trenches, or by using horizontal drilling methods to obtain the most efficient air flow pattern.

Nutrients may be supplied by blending them into water solution and then sprinkling or pumping the water across the zone to be treated. Percolation then distributes the nutrients to the affected depths for biotreatment. The water that is added supplies a method to carry the nutrients into the contaminated zone, and supplies the free water necessary for microbiological metabolism. The soil air should be maintained at or close to the dew point for optimum results. The soil moisture content should average approximately 50% of the field capacity (saturation value under unconfined conditions); if the soil porosity is 30%, then a 15% moisture content by weight would be considered appropriate.

For soil textures tighter than sands, an infiltration gallery may be installed over the area to be treated. This usually consists of a gravel bed with wellscreen distributed within it. Water and nutrients are pumped through the piping into the gravel bed so that gravity percolation may distribute the water and nutrients as evenly as possible.

The nutrient/water injection rate must be selected with care to prevent the water from carrying the contaminants down to the water table and spreading the problem into the saturated zone. Groundwater monitoring will be required to demonstrate that this has not happened, as well as to determine whether any nitrate which may have been added as a nutrient (or converted from the ammoniacal form to nitrate during microbial nitrification) has found its way to the water table.

Process Monitoring

Process monitoring is commonly performed on the effluent air stream. Hydrocarbon (or other contaminant) content is measured using one of a variety of means ranging from general methods (such as flame ionization or photoionization detectors) to gas chromatography on whole air samples or from sorbent tubes. Soil samples from the contaminated zone should be collected and analyzed after the soil effluent air stream contaminant concentration falls consistently below some preset value.

During the operation of the system, one guide to the progress of remediation is oxygen and/or carbon dioxide concentrations in the extracted air stream (or in piezometers placed in the soil at various depths). The air flow rate may be altered as required to obtain the highest amount of contaminant degradation per operating horsepower — a significant (if not one of the largest components of) cost. In some cases the effluent air stream oxygen concentration may be used as a control parameter by which to operate the system. One of the authors has personal experience in operating a bioventing system to a 5% oxygen concentration in the soil air, as a way of determining how often to run the vacuum pump.

It may be advantageous when using these methods to start and stop the vacuum system periodically. While no known kinetic data are available to compare continuous running with intermittent operation, the soil desorption rate for contaminants during shutdowns of the vacuum system provides an indication of the progress of the remediation. As the concentration of any remaining contaminants in soil air during periods with the system off diminishes, an estimate of when to conduct soil sampling can be made with improved accuracy.

Soil samples should also be periodically taken for nutrient analysis and water content. These samples need to be obtained from representative areas of the treatment zone. Nutrient and water additions may be controlled based on the test results. If nutrients and/or water is not available on-site, this service can be provided by businesses that own fertilizer trucks, tankers, and associated gear.[2]

BIOSPARGING

A second adaptation for bioventing of the subsurface is air sparging below the water table. This approach involves the injection of air below the water table in an attempt to (1) remove volatile contaminants and (2) to provide atmospheric oxygen to enhance natural biodegradation. Narrow well points are installed below the soil plume in the groundwater, and air is gently (gently!) injected into the points to strip the compounds and move them into the soil air above (see Figure 11-2). In essence, the method operates like a subsurface air stripper on the groundwater and adds oxygen to the water in the process. It was originally believed that the air formed discrete bubbles which floated up through the groundwater, but current (1997) thinking is that the air opens channels in the soil/water matrix and strips the compounds directly out of the water.

The result of this is an infusion of oxygen and volatile compounds into the vadose zone, where a vapor extraction system can be installed to remove the infused VOC. This constitutes bioventing from below. Above the water table, the system behaves as a bioventing system and the vapor extraction system may not be necessary if the compounds are degraded in the subsurface.

The greatest advantage of this method in the treatment groundwater is that no water is actually pumped. Groundwater pumping and treatment systems are notoriously slow to attain the treatment standards set for groundwater (20- to 30-year projections are not uncommon), so an effective groundwater cleanup method which does not require the movement of water is economically very attractive. It should be noted that any biodegradable compounds in soils above the water table are likely to be treated simply by the infusion of oxygen into the vadose zone, but some form of control is generally required to insure that these contaminants do not reach the ambient air. Since this remediation method is not yet well characterized from a hydrogeological modeling viewpoint, a pilot test is mandatory.

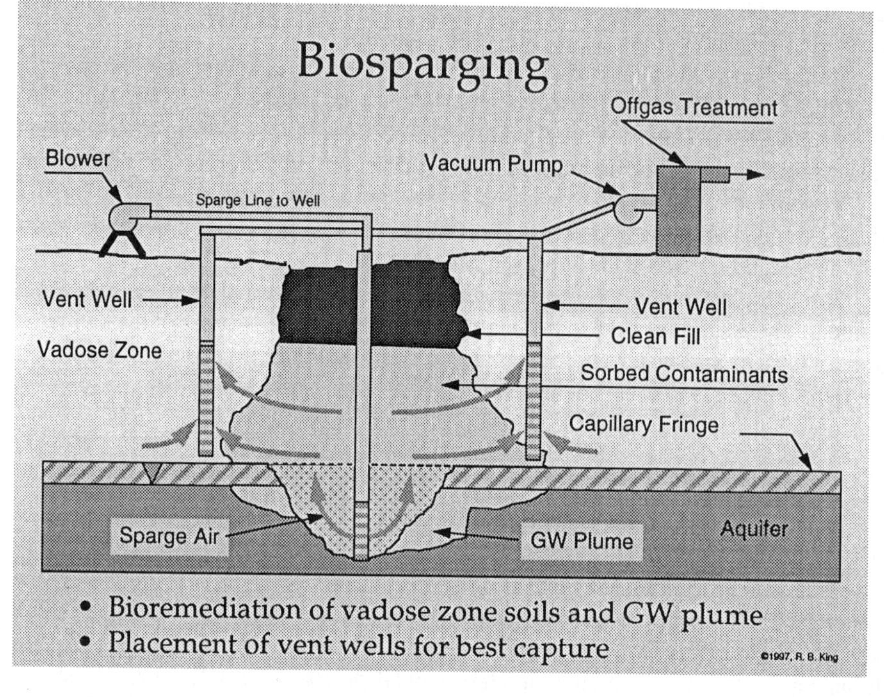

Figure 11-2 Biosparging

REFERENCES

1. Hinchee, R.E., Ong, S.K., Miller, R.N., Downey, D.C., and Frandt, R., 1992, Test Plan and Technical Protocol for a Field Treatability Test for Bioventing; Revision 2, Report to U.S. Air Force Center for Environmental Excellence, Brooks AFB, TX.
2. McMillan, D., Desert Green Bioremediation Nutrients and Equipment, Belen, NM, (505)864-2829.

CHAPTER 12

Developing Bioremediation Technologies: The Road Ahead

The first edition of this book (written during 1991) gave a few calculated insights into what the authors thought would result from research and development activities that were ongoing at that time. Since then, as a result of developments that have transpired both in the lab and field, there have been dramatic advances in some areas of applied bioremediation. Technical advances have been made in our understanding of what environmental pollutants are biodegradable, how to stimulate previously stubborn bacteria, subsurface control of substrate availability, and providing the right conditions that will encourage naturally occurring microbes to cause chemical and physical changes in their environment that result in significant improvements from a pollution or toxic standpoint.

HOW WE GOT HERE

Specific Biostimulations

As predicted, there have been a series of advances in biostimulation technology. Throughout this second edition, there are references to newly discovered techniques for bioremediation of persistent organic compounds and detoxification of metal pollutants through stimulation of indigenous bacteria.

Cometabolism

This area of research has been very rewarding, especially where the technique has been applied to sites containing halogenated solvents and PCBs.[1] Much work remains to be done and this area of research will undoubtedly produce big dividends in the future.

Advanced Acclimations and Bacterial Enzymes

The marketplace has literally exploded with a wide variety of commercial products developed along these lines. As we learn ever more about the activities of bacteria and fungi, we are never surprised at what we find. Cell-free extracts that can be applied for environmental restoration are a reality. Workers at the University of Maryland[2] have demonstrated enzyme degradation of the pesticide Parathion™. Discoveries of new enzymes and acclimative powers are always just over the horizon in this fascinating area of environmental technology. The bacterial genome is only just now being investigated. Fewer than a dozen mapping projects have been completed and the organisms under study are human pathogens. The time will come when applied microbiologists will turn their attention to the little-known microbes that function in bioremediation (80% of which have yet to even be classified). Acclimation takes place as a function of gene derepression or induction. Many of the genes of potential importance to bioremediation are unknown at present. This area of genetic research is sure to produce great dividends when specific derepression techniques are readily available for use in the field.

Microbial Chelants and Surfactants

It has long been known that microbes produce substances called chelants that bind insoluble metals into soluble organometallic complexes. Chelants are specific to metals having a certain atomic radius and charge. They are excreted from the microbial cell into the extracellular fluid environment and bind with the metal. In this soluble form the cell can transport the chelated metal across the cell membrane and obtain the metal for cell nutrition. For toxic metals, microbes can cause insoluble metals to dissolve or to volatilize. As the metals are solubilized or become volatile, they become mobile in the soil moisture or aquifer water or they can leave the aqueous state and partition to the vapor phase. Being mobile, they can be transported out of the immediate vicinity of the microbe and free the microbes from their toxic threat. Natural bacterial siderophores can bind metals and some radionucldes, such as Th^{+3} and Pu^{+4}. This knowledge led to the attempted commercial manufacture of similar compounds for metals treatment, some of which are proving to be effective for metals solubilization.[3] Chelant technology remains to be demonstrated on a field scale.

On the other hand, we have extensive knowledge about how microbes produce extracellular surfactants that can desorb pollutants bound to soil particles. This is the reason we observe hydrocarbon spikes during *in situ* bioremediation of aquifers. Specific surfactants can be induced in subsurface microbes for remediation or mobilization of aquifer contaminants.

Genetic Engineering

Cloning and recombinant DNA have been much in the news of late. Aside from the more sensational aspects of this exciting technology, the potential for environmental improvement via manipulated bacteria does not appear to present a specter or threat to our peace and well-being. In fact, transfer of genetic material between bacteria in nature is quite common. Transfer of specific DNA fragments called *plasmids* between organisms is well known. By this method, an organism that possesses a gene coding for destruction of a specific contaminant can transfer this information to neighboring microbes that are normally unable to destroy the contaminant. We are only just now learning how to exchange and insert genes in bacteria for the benefit of mankind. Plasmid transfer technology can now transfer the ability to degrade chlorinated aliphatics to recombinant *E. coli.*[4] *Clostridium* sp. have been cloned for destruction of such recalcitrant chemicals as PCBs.[5] The three major types of cloning vectors between bacteria are plasmids, bacteriophages, and cosmids. The industrial applications we enjoy as a result of this activity include improved pharmaceuticals, induced enzyme manufacture, and special biotransformations of use in bioremediation technology. A more detailed discussion of the subject can be found elsewhere.[6]

Sequencing Treatments

When bioremediation treatment technology became popular in the early 1980s, it was often used in conjunction with or as one step in a suite of technologies for site cleanup. For instance, sometimes a bioreactor was used on the surface for produced groundwater treatment prior to discharge. Land treatment bioremediation of soils sometimes followed neutralization or other chemical and physical treatments. Further combinations of using bioremediation with other technologies, sequencing, and parallel processing were encouraged and predicted. Current practice accepts and promotes its use in wastewater treatment following UV and ozone exposure.[7] Mixed waste is amenable to treatment via sequencing bioreactors along with other waste-conditioning steps.

Vadose Zone Treatment

Recent advances in bioremediation of the unsaturated zone have produced field technology for bioventing and biosparging. Certain difficulties have yet to be mastered, but the advances should be fast in coming. One of the major obstacles for this technique is the supply of nutrients in the subsurface. Gaseous forms of nitrogen and phosphorus that are bioavailable may facilitate this type of field bioremediation. The field readiness of many aspects of this important technique remains in development.

THE ROAD AHEAD

Although the foregoing advances have enhanced the overall abilities of practitioners in this technology and constitute a trend toward greater effort in research and development, we expect that the next decade will bring dramatic improvements in bioremediation technology if the trend continues. Continued emphasis upon natural solutions at the least cost will focus sharp attention upon these technologies. The authors see no reason to believe that the bioremediation approach to environmental restoration will lose favor with site owners and the regulatory community, as long as the practitioners maintain a professional attitude toward delivering a high quality product while minimizing environmental risks and keeping costs (and profits) reasonable. The areas where we should expect great technological advances in the near-term are as follows.

Metals Bioremediation

Microbes may become useful in remediation of metals contamination in aquifers through stimulation of specific biosynthetic pathways for metabolism. Bioreactor applications may be enhanced by production of chelants that can solubilize and mobilize metals for capture and removal. Metals biotreatment has already been proven practical in aquifer remediation.[8] It is only a matter of time until this innovative technology becomes fully field-ready.

Inorganics

Nitrates can be denitrified by a consortium of naturally occurring bacteria. The technology has been proven in bioreactors and industrial applications. It remains to be developed into a field-ready technique, but at least one researcher seems to be close.[8] In like manner, sulfates seem amenable to biological reduction *in situ.* The simultaneous detoxification of metals is also likely.

Fermentations

This mode of bacterial metabolism held little promise during the last decade; too many obstacles were yet to be overcome. Recent advances, however, have indicated that these biotransformations may become more controllable and may yet yield a return in terms of field-compatible techniques. They may represent the sole microbial pathway to degradation of some classes of recalcitrant compounds. An example of the tremendous potential for use of bioremedial fermentations is for destruction of explosives waste. There are probably millions of tons of contaminated soils awaiting a cost-effective alternative to incineration. Certain bacteria may have the ability to accomplish this ticklish remediation through aerobic fermentation if controls and manipulative techniques can be worked out.[9]

Mixed Waste

Not only are naturally occurring microbes of use in destroying the organic fraction of radioactive mixed waste,[10] but they can be utilized for valence-state alterations that bring nuclides out of solution for disposal.[11] There are hundreds of tons of radioactive mixed waste in storage at our national laboratories awaiting the development of treatment protocols and the availability of effective treatment methods to facilitate disposal. It is of great national importance to find acceptable ways of handling, treatment, and disposal of these toxic and hazardous wastes at reasonable risk and cost. Perhaps bioremediation will prove very useful in this application and chances are it will be very low in cost to the waste generators and on the public.

In Situ Bioremediation

As a result of the foregoing advances in bioremediation technology, the application of *in situ* aquifer bioremediation has advanced in parallel. A cursory review of the state of the art was published in 1994.[12] Our present level of understanding is sufficient for effective treatment of petroleum fuels, solvents, and many nonhalogenated organic compounds. However, it is the *halogenated compounds, metals, explosives, and nitrates* that are likely to occupy our attention in the future. Those areas showing future promise, or in current development for halogenated solvents, include innovative *in situ* treatment technologies for the following.[13]

Intrinsic

For intrinsic biodegradation of many of the common halogenated compounds, a cometabolic substrate must also be available in the aquifer to drive the system anaerobic. This substrate may also be considered by the regulatory authority to represent an additional contaminant. It is necessary that the system goes anaerobic for appreciable destruction of the target compound through reductive dehalogenation. It will be a rare occasion when the halogenated target compounds are found to occur together with a suitable cometabolic substrate in the same location and in the same time frame.

Enhanced Anaerobic

This presents a challenge due to the system design requirement for adding a suitable cometabolic substrate. Compounds such as acetate, benzoate, methanol, and organic acids are required. Many possible anaerobic halogenated by-products can accumulate in the system. Specific pathways may exist in natural populations that can degrade these accumulations all the way to mineral under anaerobic conditions, but their existence has not been confirmed or is poorly understood at present.

Enhanced Aerobic

Another system design that can accomplish the degradation of halogenated solvents, this approach also requires a suitable cometabolic substrate for dehalogenation to occur. It requires oxygen plus an organic primary substrate such as methane, propane, phenol, or toluene in a ratio of from 5:1 to 20:1 for degradation. In some cases, this means the provision of 100 ppm of primary substrate to accomplish the destruction of only 5 ppm of solvent. A major difficulty is the provision of sufficient oxygen in a system that commonly exhibits a high oxygen demand.

Sequencing Anaerobic/Aerobic

Since many highly halogenated compounds found in groundwater do not biodegrade through aerobic pathways, some means must be engineered to begin a project under anaerobic conditions to accomplish the initial dehalogenation, then switch to aerobic conditions for the final destruction of the anaerobic by-products generated in the first phase. For instance, PCE will not undergo aerobic dehalogenation; at least we don't think so at the present state of our knowledge. It requires initial treatment via anaerobic cometabolic reductive dehalogenation. However, anaerobic degradation is known to go only part way and produces high quantities of DCE and vinyl chloride, while aerobic cometabolism can take these by-products all the way to mineral. The challenge then is to initiate the biotreatment as an anaerobic remediation and switch to aerobic when the PCE has undergone the removal of the first chlorine. The aerobic phase will presumably then complete the remediation. This has been shown to take place in laboratory microcosms, but remains to be field proven. In other words, now that we know the reaction takes place under controlled conditions in the lab, next we have to engineer a workable field application together with the necessary test and monitoring protocols, hydrologic control, and nutrient delivery system. Another approach has shown that chemical oxidation followed by aerobic, then anaerobic bioremediation will convert radionuclides to insoluble valence states.[11]

Bioaugmentation

As discussed in Chapter 1, when the native microbes will not or cannot degrade the target contaminants, the remediation specialist may elect to bring in outside help in the form of engineered or acclimated microbes. One interesting development has been the use of special bacterial strains for the aerobic cometabolic destruction of TCE in groundwater.[14] This demonstration began to change our minds about what microbes are able to do under certain conditions. The opening of such new horizons is what will propel bioremediation into the next millennium. Novel and innovative *designer microbes* such as those under present development might be able to produce the proper enzymes for destruction of the target compounds by as yet unknown biosynthetic pathways, or they might be able to complement the indigenous bacteria for more complete destruction of pollutants. However, we may never see this technology take shape unless the regulators cooperate with the researchers and allow the development and field testing of these innovative attempts toward creating a natural "organic" approach to solving some of our most vexing environmental problems.

REFERENCES

1. Unterman, R., 1990, Biological approaches for PCB destruction, in Proceedings 2nd International Conference for the Remediation of PCB Contamination, Houston, TX, pp. 175–177.
2. Payne, G.F. et al., 1989, Genetic engineering approach to treating organophosphate waste, in Biotreatment: The Use of Microorganisms in the Treatment of Hazardous Materials and Hazardous Wastes, HMCRI, Washington, D.C.
3. Bisset, W.C. et al., 1997, Acyclic Polyhydroximate Chelators for Applications to Environmental Actinide Remediation, poster, WERC/HSRC '97 Joint Conference on the Environment, Albuquerque, NM.
4. Winter, R.B., Yen, K., and Ensley, B.D., 1988, Degradation of volatile chlorinated aliphatics by recombinant *E. coli*, in Hazardous Waste Treatment by Genetically Engineered or Adapted Organisms, HMCRI, Washington, D.C.
5. Li-Hua, H., 1997, Complete Sequence Analysis of 16s rDNA Clones of a *Para* and a *Meta* Anaerobic PCB Dechlorinating *Clostridium* sp., poster, WERC/HSRC '97 Joint Conference on the Environment, Albuquerque, NM.
6. Prescott, L.M., Harley, J.P., and Klein, D.A., 1993, *Microbiology*, 2nd ed., Wm. C. Brown Publishers, Dubuque, IA, Chapter 14.
7. Kearney, P.C. et al., 1986, Coumaphos™ disposal by combined microbial and uv-ozone reactions, *J. Agric. Food Chem.*, 34, 702–706.
8. Nuttall, H.E., Lutze, W., and Barton, L., 1996, Preliminary screening results for *in situ* bioremediation, International Conference on Bioremediation of Mixed Waste, Albuquerque Technical Vocational Institute, Albuquerque, NM.
9. Mondecar, M., 1997, Remediation of TNT-Contaminated Soil by Cyanobacterial Mat, poster, WERC/HSRC '97 Joint Conference on the Environment, Albuquerque, NM.
10. Wolfram, J.H., 1996, Bioprocessing scenarios for mixed waste, in International Conference on Bioremediation of Mixed Waste, Albuquerque Technical Vocational Institute, NM.
11. Tucker, M.D. et al., 1997, Assessment of Chemical and Biological Processes for Treatment of Mixed Waste, poster, WERC/HSRC '97 Joint Conference on the Environment, Albuquerque, NM.
12. Brubaker, G.R., 1994, *In situ* bioremediation of aquifers, in *GeoEnvironment 2000: Characterization, Containment, Remediation and Performance in Environmental Geotechnics,* Acar, Y.B. et al., eds.
13. Semprini, L., 1996, Bioremediation of chlorinated solvents, in International Conference on Bioremediation, Albuquerque Technical Vocational Institute, Albuquerque, NM, July.
14. Bourquin, A. and Mosteller, D.C., 1996, Pilot demonstration: aerobic bioremediation of TCE in groundwater by augmentation with *Burkholderia cepacia* PR1301, in International Conference on Bioremediation, Albuquerque Technical Vocational Institute, Albuquerque, NM, July.

APPENDIX I

State Hydrocarbon Cleanup Standards for Soil TPH[a]

State	Gasoline	Diesel	Waste Oil
AL	<100/@	<100/@	<100/@
AK	<1,000/*	<2,000/*	<2,000/*
AZ	<100 #/@	<100 #/@	<100/#/@
AR	<1,000/*	<1,000/*	<1,000/*
CA	*/# consult state LUFT manual and local authority		
CO			
CT			
DE	<100/*	<1,000/*	<1,000/*
FL	<10	<10	—
GA	<500/*	<500/*	<500/*
HI	#	#	#
ID	<200/*/@	<2000/*/@	<100/*/@
IL	*/#	*/#	*/#
IN	<100	<100	*
IA	*	*	*
KS	<100	<100	<100
KY	*/#	#	#
LA	<100/@	<300/@	*/@
ME	<5/*	<10/*	—
MD	*/@	*<10/@	*<10/@
MA	*/@	*<5/@	*<5/@
MI	#	#	#
MN	*	*	*
MS	<100	<100	<100
MO	<500/*	<500/*	<500/*
MT	<100	<100	<100
NE	<500	<500	<500
NV	<100	<100	<100
NH	<10,000/#	<7,800/#	*
NJ	*/#	*/#	*/#
NM	<50	<100	<100
NY	*	*	#
NC	*	*	*
ND	<100/*	<100/*	—
OH	*	*	*
OK	<1000/*	<1000/*	<1000/*
OR	*	*	—
PA	<200	<200	*

State Hydrocarbon Cleanup Standards for Soil TPH[a] *(continued)*

State	Gasoline	Diesel	Waste Oil
RI			
SC	*	*	*
SD	<100/*	<100/*	<100/*
TN	<1000	<1000	<1000
TX	*/#/@	*/#/@	*/#/@
UT	<30/*	<100/*	<100/*
VT	*/#	*/#	*/#
VA	*/@	*/@	*/@
WA	<100/@	<200/@	#/@
WV	*	*	—
WI	*/#	*/#	*/#
WY	<100/*	<100/*	<100/*

Note: * = site specific; # = compound specific; @ = risk assessment option.

[a] Table indicates highest allowable concentration; actual cleanup level may be lower. The state cleanup levels for groundwater follow federal SDWA, or in some states may be more stringent.

Glossary

API Separator A process unit of American Petroleum Institute (API) approved standard design for separation of oil, floating debris, solids, and wastewater. These are massive impoundments that trap oily solids and skim floating oil from refinery-generated wastewater and stormwaters. They are normally about 30 to 40 ft deep and can cover an acre in surface area. RCRA lists API sludge and several other refinery wastes as hazardous.

Absorption Action by which liquids are taken up and retained inside a semi-solid medium.

Acclimation The process of adjustment by microbes to a specific set of environmental conditions. Period during which a microbial population is "gearing up" for metabolism and degradation of a specific mix of pollutants or a specific organic compound. This is usually expressed as a lag phase when the organisms are growing at a slow rate and utilizing other substrates in preference to the target contaminant. When a critical biomass is reached in acclimated population numbers, the log growth phase begins in which the target contaminants are degraded. During acclimation there may be temporary conditions of limited substrate or environmental inhibitions that must be overcome by enzyme induction and other genetic accommodations.

Actinomycete A common class of soil bacteria active in the production of extracellular microbial antibiotics that eliminates other bacterial strains in order to improve its own competitive advantage. They are known to degrade a variety of hydrocarbons, waxes, parafins, and even synthetic rubber.

Adsorption The physical adherence of solids or chemicals to another surface; a surface active phenomenon as opposed to an *ab*sorption, in which a liquid is taken up by a solid medium or a gas dissolves in liquid.

Aerobe A microbe capable of respiration and metabolism only in the presence of free oxygen.

Aliphatic Any of the open-chain hydrocarbon compounds including paraffins, olefins, and other straight or branching chains deriving from petroleum.

Ammonia N Nitrogen from amines (NH_2^-) or ammonia (NH_3)

Ambient Whatever exists naturally in the immediate environment.

Anaerobe A microbe capable of respiring in the absence of free oxygen. Facultative anaerobes can utilize nitrate in the absence of oxygen. Strict anaerobes cannot abide the presence of any oxygen.

Anoxic Lacking free oxygen.

Aquatic A fresh-water inland habitat (as opposed to a marine or salt-water environment).

Aquifer A porous subsurface geologic formation saturated with water. The phreatic zone containing groundwater.

Atmosphere One atm of pressure equals 14.7 $lb/in.^2$.

Autotroph A lithotrophic microbe; those which utilize inorganic carbon sources as food.

BOD Biochemical oxygen demand; a test of the amount of organic carbon available for microbial digestion. A measure of the strength of a wastewater.

Bacteria Single-cell microbial plants living singly, in pairs, or forming chains of varying length.

Bioaccumulation Biological uptake of an element or compound from solution in amounts up to its ambient concentration in the environment.

Biodegradation Breakdown of pollutants and solid wastes through the action of microbes.

Biogenic Processes or products of biologic origin.

Biomagnification Biological uptake of an element or compound through biosorption and/or bioaccumulation in amounts over and above its ambient concentration in the environment becoming a sink for that element or compound by magnifying its concentration in living cells.

Biomass The total quantity of cellular material in a microbial population.

Bioremediation The practice of reclamation or cleanup of contaminated sites through employment of microorganisms for degradation and/or destruction of the pollutants.

Biosorption External uptake of an element or compound by various selective sites associated with the outer layer of the microbial cell membrane.

Biosynthesis A general term for all the biochemical metabolic processes and pathways which synthesize cellular components necessary for maintenance of life.

"Black Box" Any phenomenon that is either secret or unknown.

Buffer The ability of some soils and solutions to absorb free radicals, acids, bases, or metal ions to render a more stable chemical environment. Can be roughly equivalent to ion exchange capacity.

CERCLA Comprehensive Environmental Response, Compensation, and Liabilities Act, often referred to as SuperFund.

COD Chemical Oxygen Demand; a test of the amount of oxygen needed for the complete non-biological oxidation of compounds (waste) in a sample. A measure of the strength of a wastewater.

cfm Cubic feet per minute of airflow.

Carcinogen Capable of causing cancer. This term is very often misleading as it can mean that a certain chemical or compound has merely been shown to cause tumors in some animals under experimental conditions (usually very high concentrations in the diet) that are unrealistic in the natural environment.

Catalyst Something that hastens chemical and biochemical reaction kinetics.

Channeling The short-circuit movement of water through a treatment train or vessel that severely diminishes overall treatment efficiency.

Chelant An organic chemical molecule that binds heavy metal ions; a sequestering agent. Metal cations bound by chelants form an organo-metallic complex that is water soluble.

Clarification A step in wastewater treatment in which floc, grit, and entrained solids are removed by gravity. Solids are settled to the bottom of a clarifier and clear water flows up and out through skimmers or launders.

Cometabolite The addition of a primary substrate in order to stimulate fortuitous degradation of a target compound (the cometabolite).

Consortium A group of different populations of microorganisms in close association that form a community structure with a certain symbiosis or interrelationship in which each population contributes to the overall welfare of the group. Can refer to the member organisms in commensalism.

Cytochromes Macromolecules that operate in electron transport during cellular metabolism.

Cytoplasm The aqueous material within a living cell.

DNA Deoxyribo Nucleic Acid; the extremely complex double-helix molecule containing the genetic information (genes) for an entire organism. Directs all cellular functions and determines heredity.

Diatomaceous earth (DE) A mineral that comes from a deposit of the calcareous skeletons of marine plankton algae (diatoms) that form sediments on the ocean bottom. The mineral is a finely divided powder and serves as an excellent filter aid and can act as a foundation for microbial attachment and growth in bioreactors.

Enzyme Any of a vast number of protein biochemicals produced by living cells which can catalyze a chemical reaction involved in biosynthesis.

Estuarine A coastal or near-shore environment characterized by waters of salinity less than that of open ocean (<3.5% salt). This occurs in bays and at mouths of rivers where ocean water is impacted by fresh water inflow.

Ex situ Out of or removed from. Refers to treatments that require removal or excavation of soils, etc.

Exogenous External to the cell, or deriving from nitrogenous metabolism.

FID Flame ionization detector for volatile vapors.

Fermentation Biodegradation of organic matter in which oxidation and reduction involves transfer of electrons between organic chemical species.

Flocculation Formation of floc that can settle by gravity.

Flow model A computer code that calculates predicted flow paths and velocities for groundwater.

Friable Crumbles easily to the touch; comes apart.

Fungi A group of nonphotosynthetic plant-like, usually parasitic, organisms inhabiting the soil which are able to ferment and biodegrade many compounds and components of solid waste.

gpm Gallons per minute of flow

GC Gas chromatograph(y); an analytical instrument that can help identify individual organic chemicals by means of a gas phase separation technique.

GC/MS A highly sophisticated analytical instrument or procedure that combines a gas chromatograph with a mass spectrometer for precise identification of chemical compounds.

Gene induction Process by which microbes can induce gene expression for degradation of a specific food substrate; gene de-repression.

Geofelt A heavy synthetic felt material that allows water percolation while preventing soil movement.

Gradient The direction of groundwater flow.

Halobacteria Bacteria able to grow only in hypersaline environments.

Halotolerant Organisms able to metabolize at higher salt concentrations, but not requiring them.

Heterotroph Organisms which utilize organic food sources; organotrophs.

Hydrolysis Cleavage of organic carbon bonds where water is used in the reaction.

Hydrophobic Lacking an affinity for water.

Hydrostatic Pressure exerted by a column of water.

Hypersaline Solutions having elevated salt concentrations. Thalassic solutions are those having salts in the proportions found in seawater. Athalassic solutions have salts in altered proportions to those in seawater.

Indigenous Living naturally in an environment; naturally occuring.

in situ In place; refers to treatment of soils and groundwater where they are found without movement or excavation.

Kinetics The dynamics of motion or the speed of reaction.

Kjeldahl N A test for total nitrogen content without regard to chemical compounds containing nitrogen.

Land Ban The regulations under RCRA that restrict disposal of certain classes of waste in or on the land.

Ligand An unspecified organic chemical species.

Lyse Any condition that results in dissolving the cell membrane. Can be induced by detergent action.

Limiting factor That nutritional element or compound in least abundance without which metabolism cannot occur.

Line snubber Pressure damping device consisting of a back-pressure diaphram for smoothing pulses and providing even distribution in a flow line. Used with a diaphram or peristaltic pump.

Lithostatic Pressure exerted by a rock column; overburden.

Lithotroph Organisms which utilize inorganic carbon sources as food; autotroph.

Load capacity A measure of soil compressive strength.

Lysis Breaking up of the microbial cell and release of the cytoplasm; bursting of the cell membrane causing death.

MPN Most probable number test for bacterial count.

Marine A saltwater ocean habitat (as opposed to an aquatic or freshwater environment).

Matrix The natural material into which a pollutant or contaminant is introduced, spilled, or disposed. Usually taken to be the geochemical environment created or found in soil or water at a remediation site.

Metabolism Internal cellular biochemical reactions necessary for the maintenance of life.

Mesophilic The preference of a microbe for medium temperatures of 20 to 45°C.

Methanogen A bacteria capable of the complete reduction of carbon to methane (CH_4). They are strict anaerobes requiring a reducing environment.

Methanotroph A bacteria capable of utilizing methane as a substrate. Complete oxidation of methane to carbon dioxide occurs by aerobic degradation pathways.

Microaerophile An aerobic microbe capable of metabolism at extremely low oxygen concentrations (sometimes in environments where oxygen is below detection limit).

Microbe Microscopic unicellular life form; microorganism. Generally the vegetative form of bacteria and fungi.

Microcosm A test method utilizing soil trays, shake flasks, vials, etc., allowing maintenance of cultures in many differing regimes of environmental conditions. Allows analysis for optimum treatment conditions.

Microhabitat The immediate environment in intimate contact with living microbes which may be different from the surrounding medium.

Micronutrient A trace element essential in nutrition.

Microtox® A registered trademark indicating a specific method of testing for microbial toxicity.

Mixed waste More properly, radioactive mixed waste. A specific class of hazardous waste that comprises all mixtures of AEA regulated radioactive nuclides with any of the RCRA hazardous wastes. The RCRA waste is regulated under EPA and the RAD waste comes under the NRC or DOE regulations.

Molarity Gram molecular weight equivalents. A solution is said to be 2 *M* if it contains 2 molecular weights of a solute in a liter of solution.

Morphology Cellular shape, dimensions, and spacial arrangement.

Motile Capable of locomotion. Microbes having exterior flagella or other means of self-propulsion.

Mutagenesis Forces causing the process of mutation.

Mutation Any of a number of events which cause breaks, alter the physical sequence, or destroy portions of paired DNA chains (the nucleic acid genetic material) within a cell resulting in deletion of genetic information and altered metabolism in subsequent generations.

NAPL Nonaqueous phase liquid or free hydrocarbon.

NPDES National Pollutant Discharge Elimination System under the Clean Water Act. Applies to permits for discharge of water effluents into surface waters.

niche That position held by a species of organism within its community and defined by its food sources, its physical/chemical environment, and relationship to other organisms in the community. Competition results when two or more species vie for the same niche in a community of organisms.

OSHA Occupational Safety and Health Administration (U.S.), or the OSHA Act.

Olefinic Any open-chain unsaturated hydrocarbon containing double or triple bonded carbon atoms.

Oligomer A degradation product polymer of varying length, usually shorter than the parent polymer.

Organotroph A microbe capable of utilizing organic carbon sources as food.

Osmotic A pressure differential favoring liquid flow across a membrane from areas of greater concentration to areas of lesser concentration of water.

PAH (PNA) Polycyclic Aromatic Hydrocarbon, or Poly Nuclear Aromatic compound (equivalent terms).

PCB Polychlorinated biphenyl. A compound in common use for several years (production banned in 1979) that was developed and used for dielectric fluid in electrical switch gear and transformers; also used as heat transfer fluid and turbine lube.

PID Photo ionization detector for volatile vapors.

POTW Publicly Owned Treatment Works; refers to a wastewater treatment plant owned by a government entity.

PSH Phase-separated hydrocarbon.

Paraffinic A straight-chain saturated hydrocarbon.

Pathogen A disease-causing microbe.

Phreatic zone The saturated soil zone below the water table.

Phytoremediation Bioremediation utilizing photosynthetic plants.

Piezometer An extremely sensitive device which can measure subtle changes in air or water pressures in the subsurface.

Pleomorphic Capable of a multitude of shapes; morphologically plastic.

Plasmid A fragment of genetic material that can be passed from one microbe to another.

Polishing The last in a sequence of treatment steps just prior to discharge. The gross amount of contamination is removed ahead of the polishing step which produces a finished effluent that meets or exceeds regulatory limits.

Pore volume That volume of water occupying the pore space within a porous geologic formation.

Porosity The percent of soil pore space in natural settings through which air or water can be induced to flow.

Pretreatment Any treatment placed ahead of a wastewater treatment plant or receiving stream or sewer.

Primary tmt Treatment for removal of solids usually by gravity.

Protocol A highly specific sampling or analytic procedure.

Psychrophilic The preference of a microbe for cooler temperatures of 0 to 20°C. Some grow well at temperatures below freezing.

RCRA Resource Conservation and Recovery Act. Sets rules for all aspects of hazardous waste.

Radiolytic Breakdown of chemical structure caused by absorbed dose of radioactive emissions.

Radiotracer/radiolabeling A carbon compound that has been labeled with ^{14}C (a radioactive isotope of carbon) for the purpose of tracing the degradation pathway of the labeled compound.

RAD Radiation absorbed dose; a measure of the energy in ergs/gram of absorbed ionizing energy per unit mass of irradiated material (100 erg/g = 1 Rad and 100 RAD = 1 Gray).

Recalcitrant Any organic compound that cannot readily be degraded by microorganisms.

Refractory (see recalcitrant)

Redox Oxidation-reduction potential of a system or environment; the key condition that determines whether reactions will follow aerobic or anaerobic pathways.

REM Roentgen equivalent mammal (or man); a unit of biological dose equivalent of radiation where the number of rems is equal to the number of rads multiplied by the rbe (roentgen biological equivalent) of the given radiation for a specified effect (100 REM = 1 Seivert).

REP Roentgen equivalent physical; obsolete term. In general, the dose of any ionizing radiation which results in the absorption of about 97 erg per gram of soft tissue; essentially 1 rad.

Respirometer Laboratory device that measures the rate of oxygen uptake and/or carbon dioxide production for an indirect measurement of microbial activity and contaminant destruction.

SVE Soil vapor extraction system for *in situ* removal of volatile chemical contamination in unsaturated soils.

Secondary tmt Treatment for removal of dissolved organics in wastewater.

Sludge Any solid accumulation resulting from wastewater treatment removable by settling or filtration.

Silage Livestock feedstuffs generated by fermentation of grain (as in a silo).

SARA SuperFund Amendments and Reauthorization Act.

Statute of limitations The time limit for initiation of legal action.

Selection The ecological process by which more successful organisms replace less-adapted species through competition. Natural selection among living organisms can be thought of as an adjustment in the community to changing natural environments. As conditions change, so does the population makeup of the community. Artificial selection (and selective breeding) is the intentional alteration of the environment in order to encourage those changes that suit a specific need. If a microbial population rich in specific degraders is desired, then the environment can be altered by addition of a specific substrate which will favor those organisms able to degrade that compound.

Siderophore Highly specific iron chelating organic ligands produced by some bacteria; a siderochrome.

Sporulation The protective mechanism of endospore formation by microbes when exposed to extreme environments or starvation.

Substrate A source of carbon; any food source in microbial nutrition. A primary substrate is one that provides the immediate source of carbon for nutrition as in the case where cometabolism of a target compound (or secondary substrate) can be stimulated by addition of the primary substrate.

Surfactant A surface-active substance that can be excreted by microbial cells that can increase the solubility of organics in water making them bioavailable to the microbes.

TCLP Toxic characteristic leaching procedure (EPA). The testing procedure for the RCRA toxic waste characteristic in solid waste prior to disposal.

TOC Total organic carbon test.

TPH Total recoverable petroleum hydrocarbons (TRPH) test performed under EPA Protocol 418.1.

TSS Total suspended solids test.

Tertiary tmt Treatment for residual contaminants prior to discharge. Sometimes referred to as advanced wastewater treatment.

Thermophilic The preference of a microbe for higher temperatures above 55°C. Some grow well at temperatures well above boiling.

Transmissivity A measure of the amount of water that can be passed through a given cross-section of the subsurface; expressed as flow volume per unit area.

Transport Any movement of a fluid or particle within or through a geologic formation.

Transuranic (TRU; trans uranium) any of the metallic elements of greater atomic number than uranium (92).

Turnkey A complete system or operation that requires nothing else for proper completion of a project.

Vadose zone The unsaturated soil zone above the water table.

Valence The bonding potential for a given redox state of an atom of any element.

Viability That quality of microorganisms that allows them to grow and multiply and overcome environmental constraints. Nonviable or dormant organisms may have the potential to become viable under a set of more favorable conditions.

Well screen That portion of a well casing that is perforated to allow free flow of air or water into or out of the well.

Xenobiotic Generally, a man-made chemical, or one that does not occur in nature as a result of biological activity.

Index

E

F

G

H

I

L

M

U

V

W

X